T0113212

NAME YOUR BABY BY THE NUMBERS!
THEN DISCOVER THE SIGNIFICANCE OF
YOUR OWN NAME AND FAMILY AND FRIENDS' NAMES.

- The name Bob comes from the Old English and means bright flame. It is a #1, which suggests a person who is dynamic, assured, a natural leader.
- Kelly comes from the Gaelic and is a #2, harmonious, imaginative, and artistic.
- The master number 11 is a higher form of 2, suggesting someone highly intuitive and psychic, unified in body and soul. Jesus is 11/2.
- Oprah, from the Hebrew, meaning young deer, is the perfect #4, a builder and an organizer.

From A to Z, 1 to 9, 11, 22, 33, you'll find the full spectrum of names, personalities, and paths of destiny, including simple instructions on how to reduce any name or date to its primary number, and the significant characteristics of each. Try it, you'll love it! It's fascinating, it's enlightening. And it's fun!

NUMEROLOGY
FOR
BABY NAMES

nUMEROLOGy
for
BABY nAMES

Phyllis Vega

A Dell Book

Published by
Dell Publishing
a division of
Bantam Doubleday Dell Publishing Group, Inc.
1540 Broadway
New York, New York 10036

ISBN: 978-0-440-61390-9

BVG 01
146673257

To the memory of my once and future sister,
Elizabeth Kurke,
1946—1997.

And to my wonderful daughters,
Debbie Vega, who gave me the idea for this book, and
Sharon Garcia, who encouraged me to get it started.
And to my good friend,
Trish Macgregor, who inspired me to get it finished.

Acknowledgments

I wish to thank those who have contributed in some way to the birth, growth, and publication of this book. My agent, Al Zuckerman, and my editor, Maggie Crawford, for their faith in this project. Anne Marie O'Farrell Stack and Rob MacGregor for their helpful suggestions and practical ideas. My friends Marla Franks, Cathy Amador, Ann Maschin, Beth Cunningham, Patricia Menicci, Polly Chapman, Goldie Berman, and Jo Manning for their enthusiasm and encouragement. Family members Jorge Garcia, Yesenia Garcia, Alexis Garcia, Eduardo Rincon, and Serena Hoffman for their patience and moral support. And my late husband, Charles Vega, my beloved Charlie, who always believed in me and supported me in everything I chose to do.

"Who hath not own'd, with rapture-smitten frame,
The power of grace, the magic of a name?"

Pleasures of Hope
—Thomas Campbell

Contents

INTRODUCTION   1

CHAPTER 1: Choosing a Name   5

CHAPTER 2: Names From Other Places   10

CHAPTER 3: The Basis of Numerology   15

CHAPTER 4: How to Use This Book   19

CHAPTER 5: Number One   24

    Qualities of Number One

    Number One Names—Female

    Number One Names—Male

    Number One Names—Unisex

CHAPTER 6: Number Two/Eleven   41

    Qualities of Number Two

    Qualities of Number Eleven

    Numbers Two and Eleven Names—Female

    Numbers Two and Eleven Names—Male

    Numbers Two and Eleven Names—Unisex

CHAPTER 7: Number Three   **59**

    Qualities of Number Three

    Number Three Names—Female

    Number Three Names—Male

    Number Three Names—Unisex

CHAPTER 8: Number Four/Twenty-Two   **75**

    Qualities of Number Four

    Qualities of Number Twenty-Two

    Numbers Four and Twenty-Two Names—Female

    Numbers Four and Twenty-Two Names—Male

    Numbers Four and Twenty-Two Names—Unisex

CHAPTER 9: Number Five   **93**

    Qualities of Number Five

    Number Five Names—Female

    Number Five Names—Male

    Number Five Names—Unisex

CHAPTER 10: Number Six/Thirty-Three   **109**

    Qualities of Number Six

    Qualities of Number Thirty-Three

    Numbers Six and Thirty-Three Names—Female

    Numbers Six and Thirty-Three Names—Male

    Numbers Six and Thirty-Three Names—Unisex

CHAPTER 11: Number Seven   **130**

    Qualities of Number Seven

    Number Seven Names—Female

    Number Seven Names—Male

Number Seven Names—Unisex

CHAPTER 12: Number Eight   **147**

    Qualities of Number Eight

    Number Eight Names—Female

    Number Eight Names—Male

    Number Eight Names—Unisex

CHAPTER 13: Number Nine   **162**

    Qualities of Number Nine

    Number Nine Names—Female

    Number Nine Names—Male

    Number Nine Names—Unisex

APPENDIX 1   **177**

    Master List—Female Names

    Master List—Male Names

    Master List—Unisex Names

APPENDIX 2   **312**

    Lucky Colors, Gems, and Metals Chart

APPENDIX 3   **314**

    Worksheets

Introduction

The world knows you by your name. It is the vehicle by which you are introduced, identified, and remembered. A name is one of the first and most important gifts a parent bestows upon a newborn child. Think about it: How many times during the course of your lifetime have you said, "My name is . . ."? In choosing a name for your baby you are making a major decision that he or she will have to live with for a long time. This book has been designed to help you make that all-important choice in a new, interesting, and different way.

Numerologists believe that a name is made up of a series of numerical vibrations that contain the essence of a person's identity. A chart of number/letter correspondences is used to compute the numerological value of a name and thus determine its vibrational qualities.

Many people do not believe in numerology and consider it, along with astrology and the tarot, as unscientific hocus pocus. Others are "true believers" willing to accept anything that will tell them what they want to hear. My own view is that systems of divination are tools that can help us focus our inner senses. They stimulate natural intuition by opening up channels of information not otherwise available to the conscious mind.

It doesn't matter so much *how* or *why* the system works. What really matters is whether its techniques can provide you with meaningful information and valid answers to your questions. I cannot persuade you that numerology works simply by explaining it and its history to you. The truth of it is something that you must discover for yourself.

The ancients believed that a person's name mystically encoded information about his essential character, personality, and destiny. In this book you will learn how to select a name that has a numerological value associated with personality characteristics that you admire, and which you believe will give your child a head start and be an asset throughout life.

Where several possibilities exist for the meanings and languages of origin of a name, I have relied on my own judgment. For language of origin I have consistently chosen the first or earliest language in which the name appeared. Shortened forms and nicknames are given the same meanings as the names from which they are derived. The classification of a name as female, male, or unisex is something that is always changing, and I followed current usage. With regard to personal pronouns, I have used he and she interchangeably in order to avoid the awkwardness of using he/she.

It is my hope that this book will not only be used for naming babies. I trust that while reading it you will want to look up the meanings and personality characteristics of your own names and those of your friends and family. The answers you find will empower you because they lead to a deeper understanding of who you are.

I wish you the best of luck in your search for the perfect,

most meaningful name for your beautiful new baby boy or girl. I trust that Numerology for Baby Names will make this search especially fun and interesting, and ultimately successful.

1

Choosing a name

The Right name?

If given a choice, most people would probably not pick the names they were given at birth. Since the popularity of a name is subject to change from generation to generation, it is likely that the fashionable names given to a child today will be out of style by the time he is an adult.

No matter how much thought you give to the selection of your baby's name, there is no way to guarantee that it will be pleasing to your offspring in the years to come. I wonder how many conservative children of "hippie" parents are happy with their unique and unusual names. Certain names never go out of style, but even the most popular traditional names are not agreeable to everyone. Since it is impossible to predict what name a child will prefer, parents should select the names that *they* like best.

This book will assist you by providing information about names and numerology that will help you make your choice. But in the end the decision is your own, and whenever you are in doubt you are urged to follow the prompting of your heart.

First names First

We are living in a country where first names are used
most of the time. Since your child is likely to be referred to
almost exclusively by his first name, a good deal of thought
should be given to its selection.

In modern numerology, as in our society, the first name is
considered the most important. Surnames are connected to
family and shared with parents and siblings, but first names
are personal and more closely connected to individual identity.

Middle names

Middle names are important. Three names have become
the standard in our world. Every form you fill out has a
place for a middle name. Living as we do in a computerized
society a middle name becomes a necessity for identifica-
tion purposes.

Another reason for providing your child with a middle
name is that it gives another option to those who decide
they do not care to be called by their first name. In my
experience people who have no middle name often feel
cheated, and it is not uncommon for them to pick a middle
name for themselves.

Surnames

The names you choose should sound pleasing when spoken together with your surname. Though your child will
most often be called by her first name, it is important that
this name combine well with your family name. The best
test of sound is the "ear" test. Say aloud with your last
name each of the names you are considering. The sounds
should be harmonious and the rhythm pleasing to the ear.

Initials

Be sure to consider the words that your child's initials spell. No one wants PIG, FAT, or DIM monogrammed on a suitcase. It is also important to make sure that the initials are not an acronym for a nickname. Unless your child's name is Robert, it would be unfortunate to have the initials *BOB*.

Unisex Names

Unisex names are becoming more common. There is a trend, strongly influenced by television, to give boys' names to girls and to a lesser extent girls' names to boys.

Although some people are still touchy about the fact that gender cannot always be determined from a person's first name, I believe that unisex names are the wave of the future. Perhaps in the twenty-first century the list of unisex names will be as long as those of male and female names.

Names From the Arts

The entertainment industry is on the cutting edge with regard to trends or vogues in names. Movies, television, literature, and sports have a tremendous effect on what we name our children. More than any other medium, television has helped create trends in names. Most pervasive in influencing naming are soap operas, where the characters have a very high percentage of unusual or unisex names.

Created Names

Parents have always created names for their children. The most common way of creating a name is by combining syllables from two or more names. Thus, Mary and Joseph become Marjo. Another way of doing this is by combining two names into one. Derwin is a contemporary blend of Derek and Irwin. Created names have the advantage of being unique, but they are often changed by their owners to something more common, especially during the teen years.

Ethnic names

It is wonderful to remember your roots or to honor another tradition that you admire when naming a child. However, there are a couple of things to keep in mind when choosing unusual names. Combining different nationalities may sound exotic or comical. Be sure to avoid combinations that sound ridiculous. Like unisex names, multicultural names will probably become more common in the twenty-first century.

Religion and Tradition

Religion and tradition play a large part in the way children are named. The Catholic religion requires that its adherents use the name of a saint for the first or middle name. Christians everywhere have always favored biblical names and names implying moral virtue and religious worthiness.

Among Orthodox and Conservative Ashkenazic Jews the practice is to name a child after a relative who has passed on. However, Reform and Sephardic Jews often name a baby for a living relative.

Moslem names are usually derived from those of the Prophet, or members of his immediate family. The name of the Prophet Muhammad, with its estimated five hundred variations, is considered the most popular name in the world.

name Changes

The ancients believed that the name mystically encoded the essential character, personality, and destiny of a person.

Like the natal chart in astrology, your child's original name will always remain in the background coloring his or her entire life. When doing readings many numerologists use both the birth name and the present name. No matter how many times your child may change his name, according to the theories of numerology, he or she will always feel . the vibrational influences of the name that is on the birth certificate.

You Cannot Please Everyone

Many people have a good idea what names they would prefer for their children. However, sometimes one parent is taken completely by surprise by the other parent or by grandparents who have also picked out names for the little darlings. Choosing baby's name can turn into a tug-of-war. Although it is wonderful to get input from relatives and friends, you simply cannot please everyone, and in the end the decision must be made and agreed upon by the child's parents.

2

names From Other Places

African customs for naming a baby vary greatly, but certain kinds of names are found among many of the tribal cultures. Those most in use denote order of birth, time of birth, physical characteristics, and recent family incidents. Other typical African names such as Shark, which means "God's child," are inspired by the gods and spirits.

American Indian names are often created for the individual and they may relate to something unique about the child, like Taima: "crash of thunder," for a baby born during a storm, or to a dream one of the baby's parents had. Because of a strong belief in each person's individuality, traditional American Indian parents seldom give their own names to children. However, they might choose a name that honors a proud event in the life of an ancestor.

Arabic names have in general remained unchanged for two thousand years. Many popular Arabic names like Kamilah, which means "perfect," come from the ninety-nine qualities of God listed in the Koran. Naming customs vary from country to country. In some parts of North Africa it is customary to use the father's first name for the baby's middle name, regardless of the child's sex.

Many of the most popular names used today in Britain and the United States originated in the British Isles. Old English and Irish names along with those derived from Celtic (Briana: "strength with virtue and honor"), Gaelic (Dillon: "faithful"), and Scottish Gaelic (Douglas: "from the dark water") are very popular with English-speaking people on both sides of the Atlantic. French, Latin, German, and Danish-speaking invaders also left behind a legacy of names that have been perennial favorites in the English-speaking countries such as Russell: "red-haired," Sylvia: "from the forest," Richard: "wealthy, powerful," and Astrid: "divine strength."

Chinese parents tend to create an original name for each child. In China it is not unusual for all the names in a family to be similar and complement one another, thus Graceful Willow's sisters might be named Graceful Pearl and Graceful Flower. Because the names are so individualized, there are few common Chinese names. Because of the extreme difficulty in transliterating Chinese names into English words, very few Chinese names are currently in use in the United States. New immigrants tend to choose alternative English names for themselves and their children. In olden times parents sometimes selected ugly names or gave boys girls' names in order to fool the demons into thinking that a child was not worth harming.

In France the choice of names is often determined by social class. Traditional names that have stood the test of time, such as Isabel: "Consecrated to God," and Alain: "Handsome, cheerful," are still favored by the upper classes. The farming people also prefer traditional names, but they tend to stay away from those they consider elitist. In the cities, especially among the middle class and blue-collar workers, trendy names like Fabrice: "Craftsman" are preferred. These are characterized by Anglo-American influence but are generally pronounced in the French rather than English way.

Although Old German and Teutonic names have been the source of hundreds of names now used in the English-speaking and Spanish-speaking countries of the world, the most popular names currently used in Germany are of French origin.

Many traditional Hawaiian names like Neolani: "beautiful one from heaven" were very long and had picturesque meanings. Most contemporary Hawaiian names are adaptations of English names. However, there are some very lovely short Hawaiian names such as Ola: "life" that are still in use today. Some of the prettiest names are those created by parents from an incident at birth.

Hindus believe that God is present in everything, and most Hindu names come from the many Hindu gods, who are actually manifestations of the one God. Indra, which means "God of power" is one example. Although once considered plebeian, names from nature have become very popular, and children are often named after flowers, trees, rivers, animals, and stars. Like the Chinese, some Hindus give their children ugly names in order to trick the gods into thinking that they are not worthy of notice.

In Israel Hebrew names, such as Jonina: "little dove," Orah: "light," and Ziva: "radiant," have gained favor over Yiddish names, and names from nature, place names, and American-influenced names are prevalent.

The most common Japanese names denote or imply virtue, such as Maeko: "Truth," Reiko: "Gratitude," and Shina: "Goodness." The Japanese also favor order-of-birth and number names for their children. Created names are popular for girls, and parents often play with syllables to come up with new names. Both sexes might be named for an individual in the family or in the country who has achieved something worthwhile.

In Russia the formal name given to children at birth is known as the "passport name," and it is rarely used during childhood. Nicknames, pet names, and diminutives are popular, and there are many possible nicknames for each standard name. Throughout childhood children are called by an affectionate name derived from the formal name. A little girl named Natalia: "Natal day," could be called Natasha, Natashenka, or Tasha. At the age of sixteen they are given the name known as the "patronymic," which consists of the passport name and the father's first name, which is taken as the middle name. For a boy the suffix "-ovich" or "-evich" is added to the father's name. For a girl the suffix is "-evona" or "-ovna."

Common biblical names of Greek, Latin, and Hebrew origin have always been popular in Scandinavia. Nordic names, which reflect bravery, leadership in battle, and Norse mythology are also widely used. The popular Freya: "Noble lady," refers to the goddess who was leader of the Valkyries. Combined and hyphenated names are currently in vogue, as are names borrowed from other countries, especially the United States and England. Foreign names are pronounced the Scandinavian way.

Spanish and Latino names are usually taken from the calendar of saints, and the most typical ones have religious connotations. Sometimes all the girls in one family are given the first name Maria in honor of the Virgin Mary; each sister is then given a different middle name. Three Spanish sisters, Maria Mercedes, Maria del Carmen, and Maria del Mar, would likely be called Mercedes, Carmen and Mari Mar. Because so many similar names are used, nicknames are popular. Also common are names that signify a religious event or are otherwise connected to the church, such as Milagros (miracles) or Salvator (savior). A child of either sex might be given the names of several male and female saints, thus the popularity of Jose Maria for boys and Maria Jose for girls.

3

The Basis of Numerology

Numerology is the art and science of numbers that assigns a numerical equivalent to each letter of the alphabet. Ancient writings reveal that the use of numerological systems dates back to the beginnings of recorded history. The Sumerians, Hebrews, Chaldeans, Egyptians, Greeks, and Hindus all designed numerical systems that were used to decode secret information concealed in scripture. They employed the same systems to uncover the mysterious hidden meanings that numbers can provide about a person, place, or thing.

The two major forms of numerology still in use today are the Chaldean and Pythagorean. The Chaldean technique is the older of the two. It originated in ancient Babylon, and it combines Kabbalistic name interpretation with aspects of astrology. The more recent Pythagorean method, which was devised in Greece in the sixth century, B.C., is currently favored by most numerologists and is the one used throughout this book. Either system can be utilized with equal success. Although they have distinct differences, they are alike in some ways:

1) In both systems the numbers by and large have the same meanings.
2) They both use the same method of tallying the numbers by adding all the individual numbers together from left to right and then reducing the result by adding together the individual numbers in any double-digit number until arriving at a single digit.
3) They both use names and birthdates.
4) They both believe that the number derived from the name reveals a person's character.
5) So too, in both, the number derived from the birthdate is thought to indicate a person's destiny.

PYTHAGOREAN SYSTEM

1) Letters converted are in sequence from 1–9.
2) All compound numbers (except master numbers) are reduced to single digits.
3) It gives a more materialistic meaning.
4) It is more popular and easier to use.
5) It uses the full name given at birth.

PYTHAGOREAN CONVERSION TABLE

1	2	3	4	5	6	7	8	9	11	22	33
A	B	C	D	E	F	G	H	I			
J	K	L	M	N	O	P	Q	R	(K)		
S	T	U	V	W	X	Y	Z			(V)	(X)

CHALDEAN SYSTEM

1) Letters converted are out of sequence. The numbers used are 1–8. The number 9 can be the result after the numbers are added together and it can be interpreted, but it does not represent a single letter in the conversion table.
2) Both single and compound numbers are important.

3) It gives a more mystical meaning.
4) It is less popular and more difficult to use.
5) It uses the name by which the person is best known.

CHALDEAN CONVERSION TABLE

1	2	3	4	5	6	7	8
A	B	C	D	E	U	O	F
I	K	G	M	H	V	Z	P
J	R	L	T	N	W		
Q	S	X					
Y							

Numerology incorporates the idea expressed by Pythagoras that "all things can be expressed in numerical terms, because all things are ultimately reduced to numbers." In the Pythagoreans' view of the universe, numbers were the basis of all art, science, and music. They believed that by contemplating numbers they would discover the key to the spiritual life that would lead them to God.

The philosophy of numerology is based on the belief that everything in the universe is in motion and has its own rate of vibration. All words are comprised of letters, which have numeric equivalents. It is the numeric equivalents that reveal their vibratory rate. The series of numbers in a word or a name is like a song, which is made up of the notes on the musical scale. Each number in a word or a name has its own message, but when totalled and reduced, the single digit is like the dominant chord in a song.

Metaphysicians and occultists believe that we are born in a certain place at a particular date and time in order to learn important lessons and perform preselected tasks. The vibrations of that moment combine with those of the name we are given to point us in the appropriate direction so that we may fulfill our specific life purpose. By providing clues to our innate talents, abilities, and flaws, numerology is a method of self-discovery that can help us determine who we are and where we are headed.

4

How to Use This Book

Calculating the Numbers

The numerological values of the names in this book are Individual Name Numbers, which were obtained by adding together the unreduced values of all the vowels and consonants of each name. Their calculation was based on the following chart of number/letter correspondences from the Pythagorean system of modern numerology:

1	2	3	4	5	6	7	8	9	11	22	33
A	B	C	D	E	F	G	H	I			
J	K	L	M	N	O	P	Q	R	(K)		
S	T	U	V	W	X	Y	Z			(V)	(X)

Please note that the letters "K," "V," and "X" correspond to two different numbers. For a truer result I suggest that you start with the master number when converting these letters.

I use the term "Individual Name Number" to refer to the sum total of each single name, and "Complete Name Number" to refer to the sum total of all the names of a given individual. The Complete Name Number is variously referred to in numerology as the "Destiny," "Expression," "Abilities," or "Life Mission Number."

When calculating a name in numerology, designations such as "Junior" or "III" are not used. Therefore, Charles Raymond Vega, Charles Raymond Vega, Jr., and Charles Raymond Vega III all have the same Complete Name Number.

To calculate the numerical value of your last name, use the number/letter chart printed above and a worksheet from the back of the book. Place under the name the numbers corresponding to each of the letters, and then add the numbers. All double numbers except 11, 22, and 33 are then reduced to single digits. The numbers 11, 22, and 33 are master numbers and are not reduced further.

$$V \; E \; G \; A$$
$$22+5+7+1=35$$
$$3+5=8$$

The surname Vega has a numerical value of 8. To find the Complete Name Number for my name, we must first determine the numerical value of my first name.

$$P \; H \; Y \; L \; L \; I \; S$$
$$7+8+7+3+3+9+1=38$$
$$3+8=11$$

The first name Phyllis has a numerical value of 11.

Now, to find the Complete Name Number just add the value of the first and last names.

$$PHYLLIS \; VEGA$$
$$11+8=19$$
$$1+9=10$$
$$1+0=1$$

Thus the Complete Name Number for Phyllis Vega is 1.

Once you have calculated the numerical value of a name you can turn to chapters 5–13 to learn which qualities are associated with that number.

Calculating the Destiny Number from the Birthdate

After your baby is born you may wish to calculate her Destiny Number from the date of birth. The process is an easy one:

A) Add the day number to the month number as a whole:
14th February = 14 + 2 = 16
6 + 1 = 7

B) Add the numbers of the year together:
1998 = 1 + 9 + 9 + 8 = 27
2 + 7 = 9

C) Add the first digit to the second:
7 + 9 = 16
1 + 6 = 7

The resulting single digit, in this case 7, is the Destiny Number. Now you are ready to use the information in chapters 5–13 to shed light on your baby's destiny.

Using the Numbered Name Lists

If you want your baby to have a Complete Name that adds up to the number 1, and your last name adds up to 8, then you must choose a first name from the Number 2 list of names (chapter 6). However, if you would like your baby to have a whole name that adds up to number 2, with a surname that adds up to 8, you would choose the name from the list of Number 3 names (chapter 7). In other words, you should first decide what number you want for your baby's Complete Name, and then from that number you should subtract the total of your surname.

If the desired Complete Name total is 1, and the surname value is 8, subtract 8 from 1, using 10 (1+0=10) as a numerological equivalent of 1. 10–8=2. Therefore, you should select a name in the Number 2 name list.

Middle names must also be taken into consideration.

E S T E L L E
5+1+2+5+3+3+5=24
2+4=6

The numerical value of my middle name is 6.

PHYLLIS ESTELLE VEGA
11+6+8=25
2+5=7

With the inclusion of my middle name the numerical value of my Complete Name becomes 7.

In the event that you want the Complete Name to add up to a certain number (in this example, number 1) but do not wish to substantively change chosen first and middle names, you could choose alternative spellings or variations of the same names.

P H Y L L Y S
7+8+7+3+3+7+1=36
3+6=9

E S T E L L A
5+1+2+5+3+3+1=20
2+0=2

PHYLLYS ESTELLA VEGA
9+2+8=19
1+9=10
1+0=1

Using the Master Name Lists

Another way to use this book is by looking up in the master lists the meanings and numerical values of names that you favor. If you have no specific names in mind you might browse through the master lists to see which names appeal to you. If you already have a short list from which you have to choose your baby's names, you can quickly determine their number value by using the master lists.

Whichever method you use, it is important to remember

that when choosing names using numerology the same principle always applies. Select a name that reduces to a number in keeping with the desired character of the person you are naming.

5

number One

Qualities of number One

Keywords: energetic, independent, original, creative, individualistic, strong-willed, self-reliant, ambitious, successful, inventive, assertive, courageous, daring, generous, bold, impatient, selfish, egoistic, arrogant, obstinate, determined, stubborn, dominating, manipulative.

Symbolism: The number one represents beginnings. It is the father of all the numbers and stands for unity. One can be divided into any number, leaving it unchanged. It can also be multiplied or divided by itself and remain the same. This does not happen with any other number. Masculine in nature, the number one is symbolized by the sun and the unicorn. Its element is fire, and it is related to God, Adam, the yang, creation, ego, and being.

Personality: The number one personality is dynamic and self-assured, a natural leader. Although they are good workers, Ones resent taking orders. Since they like to come first, they are competitive by nature and willing to take risks that others may avoid. Because they are independent thinkers, Ones have the drive to put original and creative ideas into operation quickly and efficiently. Ones have mental foresight and the ability to think for themselves. They usually will not take advice and can become quite stubborn when challenged. Ones have a nature that is forceful, active, and often aggressive. One is the number of application, effort, concentration, and centralization, and gives rise to a great desire for achievement. Ones are exact about details, but since they are starters not finishers, when they become bored they may drop what they are doing and move on to the next project. Ones are very determined and usually try to overcome obstacles by meeting them head on. They have a great deal of dignity and believe deeply in their own innate ability to succeed.

There is a tendency for Ones to be self-centered, arrogant, and bossy. They must learn to guard against selfishness and egoism. Sometimes those with a one personality can become so obstinate that they cannot be diverted from a disastrous course of action, no matter what the consequences.

Careers: actor, administrator, athlete, businessperson, chief executive, director, doctor, explorer, instructor, inventor, investor, lawyer, legislator, military officer, musician, politician, salesperson, scientist, supervisor, writer, union leader.

nUMBER OnE nAMES—FEMALE

Abriana (Hebrew) Mother of the multitude.
Adalia (Old German) Noble.
Adaline (Old German) Noble.
Adinah (Hebrew) Ornament.
Agnes (Greek) Pure.
Ailani (Hawaiian) High chief.
Aileen (Greek) Light, a torch.
Aislin (Gaelic) Dream or vision.
Alameda (Spanish) Poplar tree.
Alanza (Old German) Noble and eager.
Aleria (Latin) Like an eagle.
Alina (Slavic) Bright, beautiful.
Alise (Greek) Truthful.
Alix (Greek) Helper of humankind.
Allison (Greek) Truthful.
Aloha (Hawaiian) Love.
Altagracia (Spanish) Divine grace.
Alyce (Greek) Truthful.
Amala (Arabic) Beloved.
Amalia (Old German) Industrious.
Amargo (Greek) Of eternal beauty.
Amie (Latin) The beloved.
Anatola (Greek) From the east.
Aneko (Japanese) Older sister.

Angelique (Latin) Like an angel.
Arcadia (Greek) Pastoral simplicity and happiness.
Arlene (Gaelic) A pledge.
Armida (Latin) Small warrior.
Asucena (Arabic) Lily.
Aviva (Hebrew) Springtime.
Azalea (Latin) Dry; azalea flower.
Barbie (Latin) Beautiful stranger.
Becky (Hebrew) The captivator.
Berta (Old English) Bright.
Bertrade (Old English) Shining advisor.
Brittany (Latin) From England.
Bronwen (Old Welsh) White-bosomed.
Brynn (Celtic) The heights.
Caledonia (Latin) Scottish lassie.
Calliope (Greek) Beautiful voice.
Calypso (Greek) Concealer.
Cameo (Italian) Sculptured jewel.
Camille (Latin) Young ceremonial-attendant.
Caprice (Italian) Fanciful.

Carina (Latin) Dear.
Carmelita (Hebrew) God's vineyard.
Caro (Old German) Little woman born to command.
Carolina (Old German) Little woman born to command.
Ceara (Gaelic) Spear.
Chandelle (Old French) Candle.
Chessa (Slavic) At peace.
Ciana (Hebrew) God's gracious gift.
Cindy (Greek) The moon.
Clarissa (Latin-Greek) Most brilliant.
Clementine (Latin) Merciful.
Cleopatra (Greek) Famous.
Conchita (Latin) Conception.
Concordia (Latin) Harmony.
Consolata (Latin) One who consoles.
Cora (Greek) The maiden.
Coretta (Greek) The maiden.
Corine (Greek) The maiden.
Daniela (Hebrew) God is my judge.
Delphine (Greek) Calmness.
Diandre (Latin) Divine moon goddess.
Dina (Hebrew) Judged.
Dodie (Hebrew) Beloved.
Donata (Latin) The gift.
Dori (Greek) Gift.
Dottie (Greek) Gift of God.

Dulcea (Latin) Sweet and charming.
Earline (Old English) Noblewoman.
Eda (Old German) Strives.
Edith (Old English) Rich gift.
Egypt (English) From Egypt.
Elaine (Greek) Light, a torch.
Eldora (Greek) Gift of the sun.
Electra (Greek) Brilliant, shining.
Elena (Greek) Light, a torch.
Elfrida (Old English) Elfin magic.
Elinor (Old French) Light.
Elsa (Hebrew) God's promise.
Emily (Latin) Flattering.
Emogene (Latin) Image.
Erma (Latin) High-ranking person.
Ernestine (Old English) Earnest.
Etta (Old German) Little.
Eudora (Greek) Honored gift.
Eva (Hebrew) Life-giving.
Fabia (Latin) Bean grower.
Fabiola (Latin) Bean grower.
Faye (Old German) Belief in God.
Felda (Old German) From the field.
Fenella (Gaelic) White-shouldered.
Feodora (Greek) Gift of gold.
Fidelia (Latin) Faithful.

Flair (French-English) Style, verve.

Frayda (Old German) Peaceful ruler.

Freya (Old Norse) Noble lady.

Gabriela (Hebrew) Devoted to God.

Gayla (Old French) Bright and lively.

Genifer (French) Pure white wave.

Georgeanne (Greek) Landholder, farmer.

Gianna (Hebrew) God's gracious gift.

Gillian (Greek) Young in heart and mind.

Glynnis (Old Welsh) Dweller in a valley.

Grete (Latin) A pearl.

Grizelda (Old German) Grey battle maiden.

Hadrian (Latin) From Adria.

Hannah (Hebrew) Full of grace.

Heloise (Old German) Famous woman warrior.

Henrietta (French) Mistress of the household.

Hildegarde (Old German) Battle maid.

Hortensia (Latin) Gardener.

Hulda (Austrian) Gracious.

Ilana (Hebrew) Tree.

Iliana (Greek) Light, a torch.

India (English) From India.

Indira (Hindi) God of power.

Indra (Hindi) God of power.

Indria (Hindi) God of power.

Iola (Greek) Violet-colored dawn.

Iris (Greek) Rainbow.

Isha (Hebrew) Woman.

Isleen (Gaelic) Dream or vision.

Isolde (Welsh) Fair lady.

Jameelah (Arabic) Beautiful.

Jamila (Arabic) Beautiful.

Jayne (Hebrew) God's gift of grace.

Jeanelle (Hebrew) God's gift of grace.

Jerusha (Hebrew) The married one.

Jessenia (Arabic) Flower.

Jewel (Old French) A precious gem.

Joanna (Hebrew) God's gift of grace.

Joceline (Latin) Fair and just.

Joselyn (Latin) Fair and just.

Joye (Latin) Rejoicing.

Kalila (Arabic) Beloved.

Kamilah (Arabic) Perfect.

Kassie (Greek) Disbelieved by men.

Kate (Greek) Pure maiden.

Katherine (Greek) Pure maiden.

Katie (Greek) Pure maiden.

Kay (Greek) Pure maiden.

Kimberley (Old English) From the king's wood.

Kristie (Greek) Anointed.

Kyra (Greek) Enthroned.

Lacey (Greek) Cheerful.

Laina (Old English) From the narrow road.

Lalita (Sanskrit) Without guile.

Lana (Gaelic) Comely, cheerful.

Latricia (Latin) Happiness.

Leandra (Greek) Like a lioness.

Leane (Old French) The vine.

Leatrice (Hebrew-Latin) Weary bringer of joy.

Levana (Hebrew) Moon or white.

Lexie (Greek) Helper of humankind.

Leyla (Arabic) Dark as night.

Liana (French) A climbing vine.

Liberty (English) Freedom.

Lilac (Persian) Lilac flower; bluish purple.

Lindy (Spanish) Pretty.

Linnea (Scandinavian) Lime tree.

Liora (Old French) Light.

Lois (Old German) Famous warrior maid.

Lora (Latin) Crowned with laurels.

Lorenza (Latin) Crowned with laurels.

Loretta (Latin) Crowned with laurels.

Lucia (Latin) Light.

Lucinda (Latin) Light.

Lucine (Latin) Light.

Lurline (German) Siren.

Madelaine (Hebrew) Woman of Magdala.

Madelena (Hebrew) Woman of Magdala.

Madelon (Hebrew) Woman of Magdala.

Mae (Latin) Great.

Maeve (Celtic) Intoxicating.

Mahina (Hawaiian) Moon, moonlight.

Maida (Old English) Maiden.

Maire (Hebrew) Bitterness.

Mandie (Latin) Worthy of being loved.

Mare (Hebrew) Bitterness.

Marie (Hebrew) Bitterness.

Marin (Latin) From the sea.

Martita (Arabic) Lady, mistress.

Maryanne (Hebrew) Bitter-graceful.

Matsuko (Japanese) Pine tree child.

Mavis (French) Thrush.

Maximilia (Latin) Most excellent.

Melisande (Greek) Honey bee.

Melora (Latin) Make better.

Mercy (Latin) Compassion.

Mertice (Old English) Famous and pleasant.

Meryl (French) Blackbird.

Mina (Old German) Love, tender affection.

Minerva (Greek) Wisdom.

Minnie (Old German) Love, tender affection.

Mirabella (Latin) Of extraordinary beauty.

Monica (Latin) Advisor.

Morgaine (Scottish Gaelic) From the edge of the sea.

Moriah (Hebrew) God is my teacher.

Musidora (Greek) Gift of the Muses.

Nancie (Hebrew) Full of grace.

Nanice (Hebrew) Full of grace.

Natasha (Latin) Natal day.

Nedda (Old English) Rich and happy protector.

Neri (Greek) Of the sea.

Nettie (Hebrew) Full of grace.

Neysa (Greek) Pure.

Nidia (Latin) From the nest.

Noriko (Japanese) Child of ceremony.

Odelia (Hebrew) I will praise God.

Odelinda (Hebrew) I will praise God.

Ola (Hawaiian) Life, well-being.

Olia (Latin) Symbol of peace.

Olympia (Greek) Heavenly.

Ordelia (Old German) Elf's spear.

Pansie (Greek) Fragrant.

Patience (French) Endurance with fortitude.

Paulette (Latin) Little.

Peg (Latin) A pearl.

Perdita (Latin) Lost.

Petronella (Latin) Rock or stone.

Philberta (Old German) Very brilliant.

Philomela (Greek) Lover of song.

Piper (Old English) A pipe player.

Pippy (Greek) Lover of horses.

Placida (Latin) Peaceful, serene.

Placidia (Latin) Peaceful, serene.

Prudie (Latin) Foresight.

Purity (English) Purity.

Quinta (Latin) Fifth child.

Rachelle (Hebrew) Innocent as a lamb.

Rainbow (English) Rainbow.

Raymonda (Old German) Mighty or wise protector.

Rebecca (Hebrew) The captivator.

Reva (French) Riverside.

Rhoda (Latin) Dew of the sea.

Roderica (Old German) Famous ruler.

Rosanna (Greek-Hebrew) A rose, full of grace.

Roseanne (Greek-Hebrew) A rose, full of grace.

Rosemund (Old German)
Famous guardian.
Rosita (Greek) A rose.
Roxana (Persian) Brilliant.
Roxanne (Persian) Brilliant.
Sabina (Latin) Sabine woman.
Sabrina (Latin) A Princess.
Salvia (Latin) Sage herb.
Satina (French) Satin fabric.
Sela (Hebrew) A rock.
Serafina (Hebrew) Ardent.
Seraphina (Hebrew) Ardent.
Sharde (African) Honor confers
a crown.
Shari (Hebrew) Of royal status.
Sharleen (French) Little and
womanly.
Sharlene (French) Little and
womanly.
Sherrie (Old French) Beloved.
Sibilla (Greek) Prophetess.
Solange (Latin) Alone.
Sunshine (English) Bright,
cheerful.
Susie (Hebrew) Graceful lily.
Suzanne (Hebrew) Graceful lily.
Sydelle (Old French) From Saint
Denis.

Tallis (Old French) Woodland.
Tandy (Greek) Immortality.
Tertia (Latin) The third.
Tessa (Greek) The harvester.
Thirza (Hebrew) Pleasantness.
Thomasina (Hebrew) The
devoted sister.
Tilda (Old German) Powerful in
battle.
Timothea (Greek) Honoring
God.
Tyne (Old English) River.
Tyra (Scandinavian) Of Tyr, god
of battle.
Ulla (Celtic) Sea jewel.
Ulrica (Old German) Wolf-ruler.
Vera (Latin) Truth.
Yasmeen (Arabic) Jasmine
flower.
Zahara (African) Flower.
Zaira (Arabic) Dawning.
Zandra (Greek) Helper of
humankind.
Zara (Hebrew) Of royal status.
Zena (Greek) Hospitable.
Zenia (Greek) Hospitable.
Zoe (Greek) Life.
Zulima (Hebrew) Peace.

NUMBER ONE NAMES—MALE

Adam (Hebrew) Of the red earth.

Adin (Hebrew) Sensual.

Agustin (Latin) Majestic dignity.

Ahern (Gaelic) Horse owner.

Ahren (Old German) The eagle.

Akeno (Japanese) In the morning.

Alain (Gaelic) Handsome, cheerful.

Alan (Gaelic) Handsome, cheerful.

Alberto (Old English) Noble and brilliant.

Aldrich (Old English) Old wise ruler.

Alfonso (Old German) Noble and eager.

Alfred (Old English) Good counselor.

Allyn (Gaelic) Handsome, cheerful.

Alvino (Old German) Beloved by all.

Amadeus (Latin) Love of God.

Ambrose (Greek) Immortal.

Anselm (Old German) Divine warrior.

Anton (Latin) Praiseworthy, without a peer.

Arlo (Spanish) The barberry.

Arnold (Old German) Strong as an eagle.

Arvin (Old German) People's friend.

Ashby (Old Norse) From the ash tree farm.

Atwell (Old English) From the spring.

Avram (Hebrew) The lofty one is father.

Barret (Old English) Glorious raven.

Barry (Celtic) One whose intellect is sharp.

Bartram (Old English) Brilliant raven.

Beaumont (Old French) Beautiful mountain.

Beauregard (Old French) Beautiful expression.

Beecher (Old English) One who lives by oak trees.

Benjiro (Japanese) Enjoy peace.

Berl (Old English) Cupbearer.

Bertrand (Old English) Brilliant raven.

Blakeley (Old English) From the black meadow.

Bob (Old English) Bright fame.

Boniface (Latin) One who does good.

Boyd (Gaelic) Blond.

Brant (Old English) Firebrand.

Briant (Celtic) Strength with virtue and honor.

Brice (Celtic) Quick.

Brody (Gaelic) A ditch.

Buchanan (Scottish Gaelic) House of the clergy.

Buck (Old English) Buck deer.

Burgess (Old English) Citizen of the fortified town.

Burnett (Old English) Brown complexion.

Byrne (Old English) From the brook.

Cadell (Celtic) Battle spirit.

Campbell (Scottish Gaelic) Crooked mouth.

Canute (Old Norse) Knot.

Cassius (Latin) Vain.

Cesar (Latin) Long-haired, emperor.

Chancellor (Old English) King's counsellor.

Charlton (Old English) Farmer's town.

Chuck (Old German) Strong and masculine.

Claude (Latin) Lame.

Cleary (Gaelic) The scholar.

Clifford (Old English) From the cliff-ford.

Colter (Old English) The colt herder.

Colum (Latin) Dove.

Conrad (Old German) Bold counselor.

Corwin (Old French) Friend of the heart.

Crosby (Scandinavian) From the shrine of the cross.

Curtiss (Old French) Courteous.

Damien (Greek) Constant.

Dan (Hebrew) God is my judge.

Danby (Old Norse) From the Danish settlement.

Dare (Gaelic) Great.

Darrick (Scottish Gaelic) Strong, oak-hearted.

Darrin (Greek) Gift.

Davis (Old English) Son of David.

Denzo (Japanese) Discreet.

Derward (Old English) Gatekeeper.

Desi (Latin) Yearning, sorrow.

Dimitri (Greek) From the fertile land.

Dolan (Celtic) Dark, bold.

Dolf (Old German) Noble Wolf.

Dolph (Old German) Noble Wolf.

Duff (Celtic) Dark.

Eagan (Celtic) Fiery, forceful.

Eaton (Old English) From the estate on the river.

Edmond (Old English) Prosperous protector.

Edward (Old English) Prosperous guardian.

Edwin (Old English) Rich friend.

Eldridge (Old English) Old wise ruler.

Elias (Hebrew) Jehovah is God.

Elihu (Hebrew) My God is He.

Elliot (Hebrew) Jehovah is God.

Elvy (Old English) Elfin warrior.

Emerick (Old German) Industrious ruler.

Enrico (Old German) Lord of the manor.

Erick (Old Norse) Ever-powerful, ever-ruler.

Esau (Hebrew) Hairy.

Even (Gaelic) Well born young warrior.

Everard (Old English) Hardy, brave.

Ezekiel (Hebrew) Strength of God.

Falco (Latin) Falcon.

Fausto (Latin) Fortunate.

Flinn (Gaelic) Son of the red-haired man.

Flurry (Gaelic) Leader, prince.

Garron (Old German) Guards, guardian.

Gawain (Celtic) The battle hawk.

Gaylord (Old French) Lively.

Geoffry (Old German) God's divine peace.

Geordan (Hebrew) Descending.

Giff (Old English) Bold giver.

Gil (Old German) Brilliant hostage.

Gilbert (Old German) Brilliant hostage.

Giovanni (Hebrew) God is gracious.

Gordon (Old English) Round hill.

Grady (Gaelic) Noble, illustrious.

Grantland (Old English) From the great plains.

Granville (Old French) From the large town.

Greg (Latin) Watchman.

Grey (Old English) Son of the bailiff.

Griff (Old Welsh) Fierce chief.

Harman (Latin) High-ranking person.

Harris (Old English) Son of Harry.

Haywood (Old English) From the hedged forest.

Hercules (Greek) Glorious gift.

Hereward (Old English) Army guard.

Hezekiah (Hebrew) God is my strength.

Hillard (Old German) Brave warrior.
Hobart (Old German) Bright mind.
Holt (Old English) From the forest.
Humphry (Old German) Peaceful Hun.
Hywel (Celtic) Little alert one.
Ignatius (Latin) Fiery or ardent.
Ira (Hebrew) Watchful.
Irwin (Old English) Sea friend.
Israel (Hebrew) Ruling with the Lord.
Ivan (Hebrew) God's gracious gift.
Ives (Old Norse) Battle archer.
Ivor (Old Norse) Battle archer.
Jackson (Old English) Son of Jack.
Jael (Hebrew) Mountain goat.
Jamal (Arabic) Beauty.
Jed (Hebrew) Beloved of the Lord.
Jedediah (Hebrew) Beloved of the Lord.
Jediah (Hebrew) Jehovah knows.
Jerrold (Old German) Spear-mighty.
Joab (Hebrew) God is father.
Joben (Japanese) Enjoy cleanliness.
Jorge (Greek) Landholder, farmer.

Josef (Hebrew) He shall add.
Joseph (Hebrew) He shall add.
Juan (Hebrew) God's gracious gift.
Jubal (Hebrew) The ram.
Juro (Japanese) Tenth son.
Kelby (Old German) From the farm by the spring.
Kelvin (Gaelic) From the narrow river.
Keye (Gaelic) Son of the fiery one.
Kiel (Gaelic) Handsome.
Kingston (Old English) From the king's estate.
Klaus (Greek) Leader of the people.
Knox (Old English) From the hills.
Kort (Old German) Bold.
Kraig (Scottish Gaelic) From near the crag.
Kyne (Old English) The royal one.
Ladd (English) Attendant.
Lancelot (Old French) Servant, attendant.
Landan (Old English) From the long hill.
Latham (Old Norse) From the barn.
Leon (Latin) Lion.
Link (Old English) From the settlement by the pool.

Locke (Old English) From the stronghold or enclosure.

Lorne (Latin) Crowned with laurels.

Malik (Arabic) Master.

Marden (Old English) From the valley with the pool.

Marston (Old English) From the town near the marsh.

Martyn (Latin) Follower of Mars.

Melvil (Old French) From the estate of the hard worker.

Melvyn (Gaelic) Brilliant chief.

Merwin (Old English) Lover of the sea.

Mitchell (Hebrew) Like unto the Lord.

Modesto (Latin) Modest, moderate.

Montgomery (Old English) From the rich man's mountain.

Morty (Old French) Still water.

Murdoch (Scottish Gaelic) Sailor.

Namir (Arabic) Leopard.

Neale (Gaelic) The champion.

Nestor (Greek) Traveler, wisdom.

Nevin (Gaelic) Holy, worshiper of the saints.

Newton (Old English) From the new town.

Nick (Greek) The people's victory.

Nicolas (Greek) The people's victory.

Olly (Latin) Symbol of peace.

Olvan (Latin) Symbol of peace.

Orland (Old English) From the pointed land.

Orv (Old French) From the golden estate.

Osbourne (Old English) Warrior of God.

Oskar (Old English) Divine spearman.

Oxford (Old English) From the river-crossing of the oxen.

Pablo (Latin) Little.

Park (Old English) Guardian of the park.

Parsifal (Old French) Valley-piercer.

Pascual (Latin) Born at Easter or Passover.

Patricio (Latin) Wellborn, noble.

Perceval (Old French) Valley-piercer.

Pernell (Old French) Little Peter.

Perry (Old French) Pear tree.

Pete (Latin) Rock or stone.

Peter (Latin) Rock or stone.

Phillip (Greek) Lover of horses.

Pincus (Hebrew) Oracle.

Plato (Greek) Broad-shouldered.

Quentin (Latin) Fifth child.

Quillon (Latin) The sword.

Radcliffe (Old English) From the red cliff.

Rai (Japanese) Lightning, thunder.

Ralf (Old English) Swift wolf.

Ralph (Old English) Swift wolf.

Rand (Old English) Shield-wolf.

Raynor (Old German) Mighty army.

Read (Old English) Red-haired.

Redman (Old German) Protecting counselor.

Redmond (Old German) Protecting counselor.

Reede (Old English) Red-haired.

Reeve (Old English) Steward.

Renato (Latin) Reborn.

Ridley (Old English) From the red meadow.

Robb (Old English) Bright fame.

Rod (Old English) One who rides with a knight.

Roderigo (Old German) Famous ruler.

Rogan (Gaelic) Red-haired.

Rogers (Old German) Famous spearman.

Roland (Old German) From the famous land.

Ronald (Old English) Mighty and powerful.

Rudyard (Old English) From the red enclosure.

Ruprecht (Old English) Bright fame.

Rushford (Old English) From the rush ford.

Ruttger (Old German) Famous spearman.

Ruy (Old French) King.

Sachio (Japanese) Fortunately born.

Saleem (Arabic) Safe, peace.

Salvidor (Italian) Savior.

Sanderson (Old English) Son of Alexander.

Sanson (Hebrew) Like the sun.

Sawyer (Old English) Sawer of wood.

Saxon (Old English) Swordsman.

Selwin (Old English) Friend from the manor house.

Sergio (Latin) The attendant.

Seton (Old English) From the place by the sea.

Sherlock (Old English) Fair-haired.

Siegfried (Old German) Victorious protector.

Skip (Scandinavian) Shipmaster.

Sol (Hebrew) Peaceful.

Stanislaw (Slavic) Stand of glory.

Stillman (Old English) Quiet man.

Stone (Old English) Stone.

Styles (Old English) From the stiles.

Sutton (Old English) From the southern town.

Sylvester (Latin) From the forest.

Taber (Old English) Drummer.

Taite (Old English) Cheerful.

Thaddeus (Greek) Stout-hearted.

Thayne (Old English) Attendant warrior.

Thorpe (Old English) From the village.

Tiler (Old English) Maker of tiles.

Tito (Greek) Of the giants.

Tomkin (Hebrew) The devoted brothers.

Tonio (Latin) Praiseworthy, without a peer.

Tray (Welsh) Strong as iron.

Trefor (Gaelic) Wise and discreet.

Tristam (Welsh) Sorrowful.

Tristram (Welsh) Sorrowful.

True (Old English) Faithful man.

Tynam (Gaelic) Dark.

Van (Dutch) Of noble descent.

Vasilis (Greek) Magnificent knight.

Vaughn (Welsh) Small.

Vergil (Latin) Staff bearer.

Verne (Latin) Springlike, youthful.

Vinnie (Latin) Victorious.

Waldo (Old German) Ruler.

Ward (Old English) Guardian.

Weldon (Old English) From the well on the hill.

Wernher (Old German) Defending army or warrior.

Westley (Old English) From the western meadow.

Wiatt (Old French) Little warrior.

Wilhelm (Old German) Determined guardian.

Wilt (Old English) From the town by the spring.

Win (Old English) From the friendly place.

Winfield (Old English) From the friendly field.

Woody (Old English) From the passage in the woods.

Worthington (Old English) From the farmstead.

Worthy (Old English) From the farmstead.

Yules (Greek) Youthful, downy-bearded.

Yuri (Hebrew) Jehovah is my light.

Zachary (Hebrew) Jehovah has remembered.

Zane (Hebrew) God's gracious gift.

NUMBER ONE NAMES—UNISEX

Abbe (Hebrew) My father is joy.
Abbie (Hebrew) My father is joy.
Ash (Old English) Ash tree.
Blaze (Latin) Stammerer.
Bobby (Old English) Bright fame.
Brandy (Dutch) Brandy drink.
Caley (Gaelic) Thin, slender.
Case (Gaelic) Vigilant.
Cortney (Old English) From the court.
Darin (Gaelic) Great.
Devan (Old English) Defender of Devonshire.
Eden (Hebrew) Delight.
Erin (Gaelic) Peace.
Frankie (Latin) Free.
Gerry (Old German) Spear-mighty.
Hadley (Old English) From the heath.
Hilary (Latin) Cheerful.
Kody (Old English) A cushion.
Kristin (Greek) Anointed.
Loren (Latin) Crowned with laurels.
Lyndsay (Old English) From the linden tree island.

Marlon (Old English) Falcon or hawk.
Marlow (Old English) From the hill by the lake.
Meade (Old English) From the meadow.
Meredith (Old Welsh) Guardian from the sea.
Noel (French) Born at Christmas.
Padgett (French) Useful assistant.
Pat (Latin) Wellborn, noble.
Patty (Latin) Wellborn, noble.
Paule (Latin) Little.
Pepi (Hebrew) He shall add.
Phoenix (Greek) The eagle risen from the ashes.
Randi (Old English) Swift wolf.
Reagan (Gaelic) Little king.
Rickie (Old German) Wealthy, powerful.
Scottie (Old English) Scotsman.
Sky (English) Sky.
Stacey (Latin) Stable, prosperous.
Stevy (Greek) Crowned.

Tate (Old English) Cheerful.
Taylor (Old English) Tailor.
Wally (Old German) Powerful
    warrior.

6

number Two/Eleven

number Two

Keywords: cooperative, agreeable, patient, considerate, compassionate, kind, gracious, gentle, diplomatic, understanding, balanced, artistic, romantic, loving, nurturing, tactful, receptive, sympathetic, emotional, supersensitive, shy, weak, self-conscious, deceitful, petty, restless, hoarding, deceptive, insincere, moody, changeable.

Symbolism: The number two represents pairs. It is the prime feminine number and stands for partnership and polarity. Two is the number of duality and change. The feminine nature of two is symbolized by the moon and two of anything. Its element is water, and it is related to the goddess, Eve, the yin, peacemaking, good/evil, male/female, and negative/positive.

Personality: The number two personality is harmonious, imaginative, and artistic. Twos are born followers and willingly share power and responsibility. They collect and assimilate. Partnership is their forte, and they bring cooperation, balance, and harmony to relationships. As peacemakers they use tact and diplomacy to mediate tricky situations.

Good-natured and deeply intuitive, they are good listeners who make kind and sympathetic friends. Many Twos possess psychic ability. They like beauty and order in their lives and are by nature more emotional than physical. Devoted parents, Twos enjoy creating a pleasant home atmosphere for their children. Money is important to them because it represents security and comfort. They are careful with their money and possessions, but they are not cheap. The Two is romantic, faithful, and affectionate and expects loved ones to demonstrate care and devotion. Twos dislike arguments and are usually the first to kiss and make up.

There is a tendency among Twos to be possessive of people and material objects. Twos can be shy and self-conscious, and they sometimes suffer from nervousness or depression. Emotionally thin-skinned, they are often moody and easily wounded by criticism. Most Twos require constant reassurance and encouragement. When hurt or upset they can become malicious or deceitful.

Careers: accountant, artist, bookkeeper, builder, caterer, collector, cook, computer expert, dancer, diplomat, doctor, editor, homemaker, judge, mediator, nurse, nutritionist, poet, psychiatrist, psychic, psychologist, publicist, sailor, secretary, singer, statistician, restaurateur, teacher, waiter.

Qualities of Number Eleven (A master number and the higher form of two)

Keywords: idealistic, intuitive, cerebral, spiritual, clairvoyant, visionary, humanitarian, energetic, powerful, talented, mystical, prophetic, inspired, inspirational, artistic, poetic, loving, just, harmonious, peaceful, wise, psychic, Christlike, pragmatic, fanatical, zealot, superior, cynical, cold, uncaring.

Symbolism: The number eleven is a master number, and as the higher form of two it represents the body and soul working together. Eleven stands for the transition from material to spiritual partnerships. Those with the number eleven as a major vibration are thought to be old souls whose mission in life is to lead their fellows to higher levels through inspiration and the subconscious knowledge of previous existences. Eleven is the number of revelation, abstract impressions, and transcendental enlightenment.

Personality: The number eleven personality is highly intuitive and psychic. As inspired thinkers they are able to communicate on all levels, and their idealistic natures and personal magnetism draws people to them. One of the callings of Elevens is to bring light to the world through lofty ideas. They are born to help others by leading them toward a better way of life. Elevens use their skills and knowledge to promote better standards of living for the less fortunate. Some may become ministers, teachers, evangelists, or missionaries, while the more artistically inclined will follow vocations associated with the fine arts. Many musicians, writers, dancers, poets, filmmakers, and painters are Elevens. Capable of doing two tasks at once, Elevens, in many instances, handle two careers simultaneously. As inventors and originators they often discover new truths beyond accepted realities; this is especially true when they follow their intuition. Elevens have a genuine regard for humanity, and many have an important task to perform for the good of humankind.

Elevens may become so involved in their efforts to benefit the world that they forget the needs of their own families. On a personal level some Elevens may appear cold and uncaring. They seem to spend a great deal of time on their own and often seem distant and detached from their immediate environment. Sometimes the inner priestliness of the Eleven's vibration can lead to fanaticism and/or martyrdom.

Careers: artist, aviator, dancer, filmmaker, evangelist, inventor, lecturer, minister, musician, painter, poet, politician, preacher, promoter, psychologist, teacher, union leader, writer.

NUMBERS TWO AND ELEVEN NAMES—FEMALE

Adina (Hebrew) Ornament.
Afton (Old English) One from Afton.
Agatha (Greek) Good, kind.
Aisha (African) Life.
Aisleen (Gaelic) Dream or vision.
Akiko (Japanese) Light, bright.
Alana (11/2) (Gaelic) Comely, cheerful.
Aletha (Greek) Truthful.
Alfreda (Old English) Good counselor.
Allegra (Latin) Exuberantly cheerful.
Alondra (Spanish) Lark.
Alpha (Greek) First one.
Althea (Greek) Wholesome healing.
Alula (11/2) (Arabic) The first.
Amaryllis (Greek) Amaryllis flower.
Amina (Arabic) Trustworthy.
Anda (11/2) (Japanese) Meet at the field.
Andee (Latin) Womanly.
Ann (11/2) (Hebrew) Full of grace.
Antonia (Latin) Praiseworthy, without a peer.

Antonis (Latin) Praiseworthy, without a peer.
April (Latin) Forthcoming.
Arabelle (Latin) Beautiful altar.
Aracely (Latin) Altar of heaven.
Asha (11/2) (African) Life.
Audrey (Old English) Noble strength.
Aurora (Latin) Daybreak.
Beata (11/2) (Latin) Happy.
Belen (Spanish) From Bethlehem.
Belinda (Spanish) Pretty.
Bernice (Greek) Bringer of victory.
Birdie (English) Birdlike.
Blithe (Old English) Joyful.
Bridget (Celtic) Strong and mighty.
Bridie (Celtic) Strong and mighty.
Briony (Celtic) Strength with virtue and honor.
Brunetta (Old German) Brown-haired.
Calista (Greek) Most beautiful woman.
Calla (11/2) (Greek) Beautiful.
Candy (Greek) Glittering, glowing white.

Cassie (Greek) Disbelieved by men.

Catherine (Greek) Pure maiden.

Celestine (Latin) Heavenly.

Chandel (Old French) Candle.

Chrissy (Greek) Anointed.

Christina (Greek) Anointed.

Cinderella (French) Little one of the ashes.

Cinnamon (English) Cinnamon spice.

Corazon (Spanish) Heart.

Corinna (Greek) The maiden.

Dacey (Gaelic) Southerner.

Damara (Greek) Gentle.

Damaris (Greek) Gentle.

Debby (Hebrew) The bee.

Debi (Hebrew) The bee.

Denise (Greek) Wine Goddess.

Desiree (French) Longed for.

Devondra (Old English) Defender of Devonshire.

Devora (Hebrew) The bee.

Diana (Latin) Divine moon goddess.

Dione (Greek) Daughter of heaven and earth.

Domina (Latin) A noble lady.

Dora (Greek) Gift.

Doria (Greek) Gift.

Dorinda (Greek) Beautiful golden gift.

Doris (Greek) From the ocean.

Edmonda (Old English) Prosperous protector.

Edwina (Old English) Rich friend.

Eirene (Greek) Goddess of peace.

Elissa (Hebrew) God's promise.

Elita (Latin) Chosen.

Elnora (Old French) Light.

Eloise (Old German) Famous woman warrior.

Ena (11/2) (Gaelic) Little ardent one.

Erna (Old English) Eagle.

Estella (Persian) A star.

Eulalie (Greek) Sweet-spoken.

Fiala (Czech) Violet.

Fionnula (Gaelic) White-shouldered.

Florencia (Latin) Flowering or blooming.

Florida (Latin) Flowering or blooming.

Florrie (Latin) Flowering or blooming.

Francisca (Latin) Free.

Gaye (Old French) Bright and lively.

Giovanna (Hebrew) God is gracious.

Giuliana (Greek) Young in heart and mind.

Graciela (Latin) Graceful.

Gratia (Latin) Graceful.

Gretna (Latin) A pearl.

Gwendolyn (Old Welsh) White-browed.

Gypsy (Old English) Wanderer.

Hanna (Hebrew) Full of grace.

Harmonie (Latin) Harmony.

Hatty (French) Mistress of the household.

Heather (Old English) Flowering heather.

Hedwig (Old German) Strife.

Hilde (Old German) Battle maid.

Idelle (Celtic) Bountiful.

Irena (Greek) Goddess of peace.

Isabelle (Old Spanish) Consecrated to God.

Isis (Egyptian) Supreme goddess.

Isola (Latin) Isolated, a loner.

Ivana (Hebrew) God's gracious gift.

Ivy (Old English) Ivy vine.

Jacinth (Greek) Hyacinth flower.

Jade (11/2) (Spanish) Jade stone.

Jamima (Hebrew) A dove.

Jelena (Greek) Light, a torch.

Jessika (Hebrew) Wealthy.

Jetta (11/2) (English) Intensely black.

Jezebel (Hebrew) Unexalted, impure.

Joletta (Greek) Young in heart and mind.

Jorgina (Greek) Landholder, farmer.

Josceline (Latin) Fair and just.

Josephine (Hebrew) He shall add.

Juana (11/2) (Hebrew) God's gracious gift.

Kaitlyn (Greek) Pure maiden.

Kalie (Greek) Beautiful.

Kameko (Japanese) Child of the tortoise.

Kathy (Greek) Pure maiden.

Katrina (Greek) Pure maiden.

Kaya (11/2) (Japanese) Adds a place of resting.

Kiah (African) Season's beginning.

Kishi (Japanese) Happiness to the earth.

Kittie (Greek) Pure maiden.

Kristian (Greek) Anointed.

Kristina (Greek) Anointed.

Latoya (Latin) Victorious.

Leanna (Gaelic) Loving.

Lenora (Greek) Light.

Leona (Latin) Lion.

Leontyne (Latin) Lionlike.

Lesenia (Arabic) Flower.

Leta (11/2) (Latin) Gladness.

Lettice (Latin) Gladness.

Lilliane (Latin) Lily flower.

Linnet (Old French) Songbird.

Liz (Hebrew) God's promise.

Lonna (Old English) Solitary.

Lorena (Latin) Crowned with laurels.

Lorraine (Latin) Crowned with laurels.

Lotty (French) Little and womanly.

Lucille (Latin) Light.

Lurette (German) Siren.

Lynda (Spanish) Pretty.

Lynette (Old French) Linnet bird.

Maggee (Latin) A pearl.

Maisie (Latin) A Pearl.

Malka (11/2) (Hebrew) Queen.

Marcella (Latin) Follower of Mars.

Margaret (Latin) A pearl.

Margot (Latin) A pearl.

Marian (Hebrew) Bitter, graceful.

Marilyn (Hebrew) Bitterness.

Marina (Latin) From the sea.

Marise (Japanese) Infinite, endless.

Marylin (Hebrew) Bitterness.

Melantha (Greek) Lady of the night.

Melody (Greek) Song.

Michie (Japanese) Gracefully drooping flower.

Mildred (Old English) Gentle strength.

Mindy (Greek) Dark, gentle.

Modesty (Latin) Modest, moderate.

Moira (Hebrew) Bitterness.

Nada (11/2) (Russian) Hope.

Nadia (Russian) Hope.

Nadine (Russian) Hope.

Nan (11/2) (Hebrew) Full of grace.

Nani (Hawaiian) Beautiful.

Natalya (Latin) Natal day.

Neema (Swahili) Born in prosperity.

Nereida (Greek) Sea nymph.

Nerine (Greek) Sea nymph.

Nina (Spanish) Girl.

Ninetta (Spanish) Girl.

Noemi (Hebrew) Pleasant.

Norah (Greek) Light.

Odetta (Hebrew) I will praise God.

Oralia (Latin) Golden.

Parthena (Greek) Virgin.

Paulina (Latin) Little.

Penny (Greek) The weaver.

Petrina (Latin) Rock or stone.

Phyllis (11/2) (Greek) A green branch.

Pilar (Spanish) Pillar of the church.

Pollyanna (Latin-Hebrew) Little, graceful.

Pomona (Latin) Fertile.

Questa (French) Searcher.

Rachel (Hebrew) Innocent as a lamb.

Raine (Sanskrit) Queen.

Randa (Old English) Swift wolf.

Raquel (Hebrew) Innocent as a lamb.
Reina (Latin) Queen.
Rena (Hebrew) Song.
Renee (Latin) Reborn.
Rhona (Old Norse) Rough isle.
Riane (Welsh) A witch.
Risa (Latin) Laughter.
Rochella (French) From the little rock.
Roda (Latin) Dew of the sea.
Rolanda (Old German) From the famous land.
Roma (Latin) Eternal city.
Ronalda (Old English) Mighty and powerful.
Rosalind (Spanish) Beautiful rose.
Rozina (Greek) A rose.
Ruthanne (Hebrew) Compassionate, full of grace.
Ruthy (Hebrew) Compassionate.
Sade (11/2) (Hebrew) Princess.
Sadie (Hebrew) Princess.
Salina (Latin) Salty.
Salome (Hebrew) Peace.
Samala (11/2) (Hebrew) Asked of God.
Sapphire (Greek) Sapphire stone.
Sarah (Hebrew) Of royal status.
Sari (Arabic) Most noble.
Saxona (Old German) Sword-bearer.
Selena (Greek) Moon.

Sandra (Sanskrit) Moonlike.
Shanleigh (Gaelic) Child of the old hero.
Sherilyn (Old English) From the bright meadow.
Sonya (Greek) Wisdom.
Spring (Old English) Springtime.
Steffi (Greek) Crowned.
Susan (11/2) (Hebrew) Graceful lily.
Sydel (Old French) From Saint Denis.
Tabina (Arabic) Muhammad's follower.
Tameko (Japanese) Child of good.
Tangie (English) Tangerine-colored.
Tawny (English) Of a warm sandy color.
Tiffani (Greek) Appearance of God.
Tirza (Hebrew) Cypress tree.
Titania (Greek) Giant.
Tomiko (Japanese) Happiness child.
Trish (Latin) Wellborn, noble.
Twila (Old English) Woven of a double thread.
Udele (Old English) Prosperous.
Umeko (Japanese) Plum blossom child.
Ursula (Latin) Little bear.
Vana (11/2) (Greek) Butterflies.

Verena (Old German) Defender.
Violet (Latin) Violet Flower.
Vondra (Czech) Womanly, brave.
Warda (Old German) Guardian.
Wilona (Old English) Desired.

Winnie (Old German) Peaceful friend.
Winola (Old German) Gracious friend.
Zita (Latin) Little rose.
Zona (Latin) A girdle or belt.

NUMBERS TWO AND ELEVEN NAMES—MALE

Abel (11/2) (Hebrew) Breath.
Adan (11/2) (Hebrew) Of the red earth.
Adolf (Old German) Noble wolf.
Adolph (Old German) Noble wolf.
Aidan (Gaelic) Warmth of the home.
Alasdair (Greek) Helper of humankind.
Albie (Old English) Noble and brilliant.
Albin (Latin) White.
Alford (Old English) The old ford.
Alonzo (Old German) Noble and eager.
Alvyn (Old German) Beloved by all.

Ames (11/2) (Old French) Friend.
Andrew (Greek) Strong and manly.
Ara (11/2) (Arabic) Rainmaker.
Ardley (Old English) From the domestic meadow.
Aristotle (Greek) The best.
Arne (Old German) Strong as an eagle.
Arnie (Old German) Strong as an eagle.
Atherton (Old English) Dweller at the spring farm.
Aubin (Old French) The blond one.
Audric (Old English) Wise old ruler.
Axton (Old English) From the peaceful town.

Baldemar (Old German) Bold and famous prince.

Baldwin (Old German) Bold friend.

Barney (Greek) Son of prophecy.

Bartley (Old English) Bartholomew's meadow.

Beal (11/2) (Old French) Handsome.

Beau (11/2) (Old French) Handsome.

Benito (Hebrew) Son of my right hand.

Benjy (Hebrew) Son of my right hand.

Bentley (Old English) From the bent grass meadow.

Beresford (Old English) From the barley wood.

Blanco (Spanish) White, fair.

Borak (Arabic) The lightning.

Bradlee (Old English) From the broad meadow.

Braun (Old German) Brown-haired.

Brewster (Old English) Brewer.

Bryon (Celtic) Strength with virtue and honor.

Buddy (Old English) Herald, messenger.

Burdon (Old English) From the castle on the hill.

Burley (Old English) From the castle by the meadow.

Byron (Old French) From the cottage.

Cabe (11/2) (Old English) Ropemaker.

Caesar (Latin) Long-haired, emperor.

Calhoun (Celtic) From the narrow forest.

Camillo (Latin) Young ceremonial attendant.

Caradock (Celtic) Beloved.

Carlton (Old English) Farmer's town.

Carnell (Old English) Defender of the castle.

Carter (Old English) Cart driver.

Chace (Old French) Hunter.

Chaney (French) Oak wood.

Chapman (Old English) Merchant.

Chico (Spanish) Small boy.

Christian (Greek) Anointed.

Claiborn (Old English) Born of clay.

Claudio (Latin) Lame.

Claus (11/2) (Latin) Lame.

Cleve (Old English) From the cliff-land.

Collier (Old English) Charcoal merchant, miner.

Collin (Gaelic) Strong and virile.

Conan (Celtic) High and mighty.

Corbett (Latin) Raven.

Cort (Old German) Bold.
Cosmo (Greek) The universe, well-ordered.
Cowan (Gaelic) Hillside hollow.
Craig (Scottish Gaelic) From near the crag.
Crandall (Old English) From the crane valley.
Cromwell (Old English) Dweller by the winding brook.
Cross (Scandinavian) From the shrine of the cross.
Daley (Gaelic) Assembly, gathering.
Damon (Greek) Constant.
Dario (Greek) Wealthy.
Darvell (Old English) Town of eagles.
Delfino (Greek) Calmness.
Delmont (Old English) Of the mountain.
Delwyn (Old English) Proud friend.
Dennis (Greek) God of wine.
Desmond (Gaelic) Man from the south.
Donnelly (Gaelic) Brave dark man.
Doug (Scottish Gaelic) From the dark water.
Dugan (Gaelic) Dark.
Dylan (Welsh) Man from the sea.
Easton (Old English) From east town.

Eddy (Old English) Prosperous guardian.
Efrem (Hebrew) Very fruitful.
Einar (Old Norse) Warrior chief.
Elgin (Celtic) Noble, white.
Elson (Old English) Son of Elias.
Elwood (Old English) From the old forest.
Emmet (Old German) Industrious ruler.
Erling (Old English) Son of a nobleman.
Eustace (Greek) Steadfast.
Everley (Old English) Field of the wild boar.
Ewen (Gaelic) Wellborn young warrior.
Fabion (Latin) Bean grower.
Fagan (Gaelic) The fiery one.
Fairfax (Old English) Fair-haired.
Farid (Arabic) Exceptional, unequaled.
Farold (Old English) Mighty traveler.
Feliciano (Latin) Joyous, fortunate.
Felix (Latin) Joyous, fortunate.
Fenton (Old English) From the marshland farm.
Fermin (Latin) Energetic.
Flavian (Latin) Yellow-haired.
Forbes (Gaelic) Prosperous.

Forest (Old French) Forest woodsman.

Forrest (Old French) Forest woodsman.

Foster (Latin) Keeper of the woods.

Franklyn (Latin) Free.

Franz (Latin) Free.

Frazer (Old English) Curly-haired.

Fredrick (Old German) Peaceful ruler.

Fuller (Old English) One who shrinks and thickens cloth.

Gage (Old French) Pledge.

Galvin (Gaelic) Sparrow.

Gannon (Gaelic) Fair-complected.

Garey (Old English) Spear carrier.

Garrison (Old English) Spear-fortified town.

Garson (Old English) Spear-fortified town.

Garwood (Old English) From the fir forest.

Geraint (Celtic) Old.

Gerald (Old German) Spear-mighty.

Gifford (Old English) Bold giver.

Gonsales (Old German) Complete warrior.

Griffith (Old Welsh) Fierce chief.

Guido (Latin) Life.

Gus (11/2) (Swedish) Staff of the Goths.

Hamilton (Old English) From the proud estate.

Hanley (Old English) From the high pasture.

Hardy (Old German) Bold and daring.

Hart (Old English) Male deer.

Haskel (Hebrew) Understanding.

Heber (Hebrew) Partner.

Hirsch (Hebrew) Deer.

Hubert (Old German) Brilliant, shining mind.

Hume (Old English) Home.

Ilya (Hebrew) Jehovah is God.

Inglebert (Old German) Bright as an angel.

Innis (Gaelic) From the island.

Isaiah (Hebrew) God is my helper.

Isidor (Greek) Gift of Isis.

Ivanhoe (Hebrew) God's gracious gift.

Jacan (11/2) (Hebrew) Trouble.

Jaime (Hebrew) The supplanter.

Jaimie (Hebrew) The supplanter.

Jamaal (11/2) (Arabic) Beauty.

Jared (Hebrew) One who rules.

Jarrett (Old German) Spear-mighty.

Javier (Arabic) Bright.

Jedreck (Polish) A strong man.

Jeraldo (Old German) Spear-mighty.

Jerard (Old English) Spear-strong.

Jervis (Old German) Keen with a spear.

Jesus (11/2) (Hebrew) God will help.

John (Hebrew) God's gracious gift.

Jonathan (Hebrew) Jehovah gave.

Joshua (Hebrew) Jehovah saves.

Julius (Greek) Youthful, downy-bearded.

Kolby (Old English) From the dark farm.

Kynan (Celtic) High and mighty.

Larry (Latin) Crowned with laurels.

Lennon (Gaelic) Little cloak.

Lombard (Latin) Long-bearded.

London (Old English) Fortress of the moon.

Lothar (Old German) Famous warrior.

Lucas (11/2) (Latin) Light.

Lyman (Old English) A man from the meadow.

Lyonel (Old French) Young lion.

Macey (Old French) Stoneworker.

Malachi (Hebrew) Angel.

Mario (Latin) Follower of Mars.

Marley (Old English) From the hill by the lake.

Maro (Japanese) Myself.

Max (11/2) (Latin) Most excellent.

Mike (Hebrew) Like unto the Lord.

Milton (Old English) From the mill town.

Morris (Latin) Dark-skinned.

Muhammad (Arabic) Praised.

Myles (Latin) Soldier, warrior.

Nairn (Celtic) Dweller by the alder tree.

Napoleon (Greek) Lion of the woodland dell.

Nemo (Greek) From the glen.

Nevins (Gaelic) Holy, worshiper of the saints.

Nigel (Latin) Black-haired.

Noah (Hebrew) Rest, comfort.

Nolan (Gaelic) Famous or noble.

Norbert (Old Norse) Brilliant hero.

Norby (Old Norse) Brilliant hero.

Normie (Old French) A northman.

Octavius (Latin) The eighth.

Omar (Arabic) Most high follower of the Prophet.

Omarr (Arabic) Most high follower of the Prophet.

Orestes (Greek) Mountain man.

Orin (Gaelic) Pale-complected.

Orrin (Gaelic) Pale-complected.

Osborn (Old English) Warrior of God.

Oscar (Old English) Divine spearman.

Oswald (Old English) Divinely powerful.

Padraig (Latin) Wellborn, noble.

Pasquale (Latin) Born at Easter or Passover.

Pedraic (Latin) Wellborn, noble.

Phelan (Gaelic) Brave as a wolf.

Pierce (Latin) Rock or stone.

Pierrot (Latin) Rock or stone.

Porter (Latin) Door keeper.

Powell (Celtic) Alert.

Prince (Latin) Chief, prince.

Prinz (Latin) Chief, prince.

Pryor (Latin) Head of a monastery.

Quinton (Latin) Fifth child.

Rainer (Old German) Mighty army.

Ramiro (Spanish) Wise, renowned.

Rancell (African) Borrowed all.

Ranier (Old Norse) Mighty army.

Reuben (Hebrew) Behold a son.

Rex (Latin) King.

Rich (Old German) Wealthy, powerful.

Rock (Old English) From the rock.

Roden (Old English) From the reed valley.

Roderick (Old German) Famous ruler.

Rodman (Old English) One who rides with a knight.

Rohan (Gaelic) Red-haired.

Rolfe (Old English) Swift-wolf.

Romulus (Latin) Citizen of Rome.

Ron (Old English) Mighty and powerful.

Rooney (Gaelic) Red-haired.

Roswald (Old English) From a field of roses.

Royall (Old French) Royal.

Rudd (Old English) From the red enclosure.

Rufino (Latin) Red-haired.

Ruskin (Old French) Red-haired.

Rustin (Old French) Red-haired.

Rutledge (Old English) From the red pool.

Salvador (Italian) Savior.

Sanborn (Old English) From the sandy brook.

Santino (Latin) Saints.

Saunders (Old English) Son of Alexander.

Severin (Latin) Strict, restrained.

Sheffield (Old English) From the crooked field.

Shepherd (Old English) Shepherd.

Silvano (Latin) From the forest.

Skeeter (Old English) The swift.

Sly (11/2) (Latin) From the forest.

Solly (Hebrew) Peaceful.

Spark (Old English) Happy.

Stefan (Greek) Crowned.

Stephan (Greek) Crowned.

Sullivan (Gaelic) Black-eyed.

Tabor (Old English) Drummer.

Tadd (11/2) (Greek) Stout-hearted.

Ted (11/2) (Greek) Gift of God.

Thatcher (Old English) Roof thatcher.

Thom (Hebrew) The devoted brother.

Thorstein (Scandinavian) Thor's stone.

Timothy (Greek) Honoring God.

Titos (Greek) Of the giants.

Tomaso (Hebrew) The devoted brother.

Tomlin (Hebrew) The devoted brother.

Toru (Japanese) Sea.

Tremayne (Celtic) From the house of stone.

Trueman (Old English) Faithful man.

Trumaine (Old English) Faithful man.

Tynan (Gaelic) Dark.

Ulrick (Old German) Wolf-ruler.

Urbain (Latin) From the city.

Urban (Latin) From the city.

Uriel (Hebrew) God is my flame.

Vaughan (Welsh) Small.

Vincente (Latin) Victorious.

Virgilio (Latin) Staff-bearer.

Vittorio (Latin) Victorious.

Walt (11/2) (Old German) Peaceful warrior.

Ware (Old German) Watchman or defender.

Webster (Old English) Weaver.

Werner (Old German) Defending army or warrior.

Wes (11/2) (Old English) From the western meadow.

Westbrook (Old English) From the western brook.

Wiley (Old English) From the water meadow.

Wilfredo (Old German) Resolute and peaceful.

Will (Old German) Determined guardian.

Wilson (Old English) Son of Will.

Wolf (Old German) Wolf.

Wylie (Old English) From the water meadow.

Zebadiah (Hebrew) The Lord's
   gift.
Zeke (Hebrew) Strength of God.

NUMBERS TWO AND ELEVEN NAMES—UNISEX

Adrian (Latin) Black earth.
Alexie (Greek) Helper of
   humankind.
Blaire (Gaelic) From the plain.
Brett (Celtic) Native of Brittany.
Cary (Old German) Strong and
   masculine.
Chandler (Old French)
   Candlemaker.
Charlie (Old German) Strong
   and masculine.
Cody (Old English) A cushion.
Dana (11/2) (Old English) From
   Denmark.
Evelyn (Hebrew) Life-giving.
Gail (Old English) Gay, lively.
Garnet (Old English) Garnet
   gem.
Glen (Old Welsh) Dweller in a
   valley.
Hallie (Old English) Hay
   meadow.
Jamie (Hebrew) The supplanter.

Kacie (Gaelic) Vigilant.
Kelly (Gaelic) Warrior.
Kirby (Old Norse) From the
   church village.
Korey (Gaelic) From the hollow.
Lynn (Old English) From the
   waterfalls.
Maddy (Hebrew) Gift of God.
Murphy (Gaelic) Sea warrior.
Nevada (Spanish) Snow-clad.
Page (French) Useful assistant.
Paige (French) Useful assistant.
Palmer (Latin) The palm-bearer.
Randee (Old English) Swift
   wolf.
Rennie (Latin) Reborn.
Riki (Old German) Wealthy,
   powerful.
Robyn (Old English) Bright
   fame.
Shane (Hebrew) God's gracious
   gift.

Shawn (Hebrew) God's gracious gift.

Sidnee (Old French) From Saint Denis.

Stormy (English) Stormy.

Sydney (Old French) From Saint Denis.

Tony (Latin) Praiseworthy, without a peer.

Tracie (Gaelic) Battler.

Tristan (Welsh) Sorrowful.

Ventura (Spanish) Happiness, good luck.

7

number Three

Qualities of number Three

Keywords: communicative, proud, active, optimistic, joyful, creative, independent, ambitious, conscientious, popular, expressive, sociable, outgoing, youthful, cheerful, lucky, inspiring, adaptable, determined, kind, generous, versatile, unpredictable, boastful, wasteful, conceited, superficial, extravagant, scheming, dictatorial, gossipy, vain, flirtatious, petulant, sullen.

Symbolism: The number three represents trinities. Three is masculine in nature and stands for expression. Its symbols are Jupiter and the triangle, and its element is fire. Three is related to time—past, present, future—and to the family—father, mother, child. It is the number of birth, life, and death, and beginning, middle, end.

Personality: The number three personality is ambitious, self-expressive, and hard working with a drive toward success. Threes strive to rise in the world, and their keen dislike of occupying subordinate positions fuels their climb to the top. They are born entertainers, no matter what career they follow. Threes have the gift of communication and can reach people in all walks of life. They are cheerful, witty, and charming, and they are able to help others become equally light-hearted. Most Threes are extremely lucky and seem to have been born under a fortuitous star. Even their difficulties and problems have a way of working out for the best. Many Threes are interested in the occult and the unknown and may have their own ideas about religion and philosophy. In their personal lives Threes are devoted to their friends and families. They are generally loyal, generous, warm, and loving, but once they determine that a relationship has ended they will leave and not look back.

Threes may become conceited prima donnas who wish to be center stage at all times. They can often be too outspoken, and they are quick to anger if they discover someone trying to take advantage of them. Threes can be somewhat shallow and vain, especially when their inborn fear of aging comes into play.

Careers: actor, artist, administrator, corporate executive, entertainer, fashion designer, financier, jeweler, lawyer, musician, illustrator, lecturer, minister, photographer, poet, politician, president, prophet, publisher, salesperson, singer, writer, youth counselor.

NUMBER THREE NAMES—FEMALE

Adana (Hebrew) Of the red earth.

Adeliz (Greek) Truthful.

Adelle (Old German) Of noble rank.

Adolpha (Old German) Noble wolf.

Adora (Latin) Beloved.

Albina (Latin) White.

Alejandra (Greek) Helper of humankind.

Alice (Greek) Truthful.

Alizah (Hebrew) Joyful.

Alwyn (Old German) Beloved by all.

Amber (Old French) Amber jewel.

Amethyst (Greek) Against intoxication.

Amy (Latin) The beloved.

Anabella (Hebrew-Latin) Graceful, beautiful.

Analeigh (Hebrew-Old English) Graceful, meadow.

Anna (Hebrew) Full of grace.

Annabelle (Hebrew-Latin) Graceful, beautiful.

Annis (Hebrew) Full of grace.

Aquilina (Spanish) An eagle.

Aracellie (Latin) Altar of heaven.

Argenta (Latin) Silvery.

Ashlynn (Old English) From the ash tree meadow.

Ayla (Hebrew) Oak tree.

Bahira (Arabic) Dazzling.

Bathsheba (Hebrew) Daughter of the oath.

Beda (Old English) Warrior maiden.

Bee (Latin) Bringer of joy.

Benecia (Latin) Blessed.

Bethany (Aramaic) House of poverty.

Bianca (Latin) White.

Briar (Old English) A thorny plant.

Bronwyn (Old Welsh) White-bosomed.

Brunhilde (Old German) Armed battle maid.

Bryanna (Celtic) Strength with virtue and honor.

Cailin (Gaelic) Thin, slender.

Caitlyn (Greek) Pure maiden.

Calida (Spanish) Warm, ardent.

Candice (Greek) Glittering, glowing white.

Carlie (Old German) Little woman born to command.

Cathy (Greek) Pure maiden.

Cecily (Latin) Dim-sighted or blind.

Celia (Latin) Dim-sighted or blind.

Charity (Latin) Brotherly love.

Charleen (French) Little and womanly.

Charlene (French) Little and womanly.

Charlotte (French) Little and womanly.

Cherie (Old French) Beloved.

Chloris (Greek) Pale.

Christiana (Greek) Anointed.

Christy (Greek) Anointed.

Cicely (Latin) Dim-sighted or blind.

Cindi (Greek) The moon.

Claire (Latin) Bright, shining girl.

Clarebelle (French) Brilliant, beautiful.

Clea (Greek) Famous.

Clio (Greek) Famous.

Clover (Old English) Clover blossom.

Colleen (Gaelic) Girl.

Corissa (Greek) The maiden.

Crissy (Greek) Anointed.

Cristina (Greek) Anointed.

Damita (Spanish) Little noble lady.

Daphne (Greek) Laurel or bay tree.

Davita (Hebrew) Beloved.

Deanna (Latin) Divine Moon Goddess.

Debra (Hebrew) The bee.

Deitra (Greek) From the fertile land.

Deka (Somali) Pleasing.

Demelza (English) From Demelza in Cornwall.

Demetra (Greek) From the fertile land.

Demetria (Greek) From the fertile land.

Dessa (Greek) Wandering.

Devonna (Old English) Defender of Devonshire.

Diantha (Latin-Greek) Divine Moon Goddess.

Dody (Hebrew) Beloved.

Donna (Latin) Lady.

Dot (Greek) Gift of God.

Dotty (Greek) Gift of God.

Ella (Greek) Light.

Ellen (Greek) Light.

Elrica (Old German) Ruler of all.

Elyse (Hebrew) God's promise.

Enyd (Old Welsh) Purity.

Esther (Persian) A star.

Eunice (Greek) The people's victory.

Evita (Hebrew) Life-giving.

Evonne (Hebrew) Life-giving.

Fae (Old German) Belief in God.

Felicitas (Latin) Joyous, fortunate.

Fifi (Hebrew) He shall add.

Filomena (Greek) Love song.

Florina (Latin) Flowering or blooming.

Frances (Latin) Free.

Gala (Old English) Gay, lively.

Gemma (Latin) Precious stone.

Georgette (Greek) Landholder, farmer.

Geralda (Old German) Spear-mighty.

Geraldine (Old German) Spear-mighty.

Golda (Old English) Golden.

Gracia (Latin) Graceful.

Grier (Latin) Watchman.

Griselda (Old German) Grey battle maiden.

Gwyneth (Old Welsh) White-browed.

Hedwiga (Old German) Strife.

Helma (Old German) Helmet.

Hester (Persian) A star.

Hiberna (Latin) Girl from Ireland.

Honore (Latin) Honorable.

Ianthe (Greek) Violet-colored flower.

Iona (Greek) Violet-colored stone.

Isabel (Old Spanish) Consecrated to God.

Jane (Hebrew) God's gift of grace.

Janette (Hebrew) God's gift of grace.

Jarietta (Arabic) Earthen water jug.

Jasmin (Arabic) Fragrant flower.

Javiera (Arabic) Bright.

Jena (Arabic) A small bird.

Jennie (Hebrew) God's gift of grace.

Jessica (Hebrew) Wealthy.

Jinx (Latin) Charming spell.

Jocelyn (Latin) Fair and just.

Kalani (Hawaiian) The sky.

Kandace (Greek) Glittering, glowing white.

Kari (Greek) Pure maiden.

Keala (Hawaiian) The pathway.

Kerrin (Greek) Pure maiden.

Keturah (Hebrew) Sacrifice.

Kira (Old Persian) The sun.

Kirsty (Norse) Anointed.

Kourtney (Old English) From the court.

Lace (English) Ornamental trimming.

Lainey (Greek) Light, a torch.

Leila (Arabic) Dark as night.

Leonore (Greek) Light.

Leotie (American Indian) Prairie flower.

Lilian (Latin) Lily flower.

Liza (Hebrew) God's promise.
Ludmilla (Slavic) Loved by the people.
Lulu (Old German) Famous battle maid.
Luna (Latin) The moon.
Lunetta (Latin) Little moon.
Madge (Latin) A pearl.
Magdala (Hebrew) Woman of Magdala.
Magdalen (Hebrew) Woman of Magdala.
Magdalia (Hebrew) Woman of Magdala.
Mandy (Latin) Worthy of being loved.
Margareta (Latin) A pearl.
Mariana (Hebrew) Bitter, graceful.
Marianne (Hebrew) Bitter, graceful.
Marilla (Latin) Shining sea.
Marvella (Latin) Wonderful, extraordinary.
Mary (Hebrew) Bitterness.
Maryleen (Hebrew) Bitterness.
Maud (Old German) Powerful in battle.
Maxine (Latin) Most excellent.
May (Hebrew) Bitterness.
Meghan (Gaelic) Soft, gentle.
Miya (Japanese) Three arrows.
Mollie (Hebrew) Bitterness.
Morena (Spanish) Brown, brown-haired.

Morisa (Latin) Dark-skinned.
Myisha (Arabic) Woman, life.
Myra (Greek) Fragrant ointment.
Myrtle (Greek) Myrtle plant.
Nana (Hebrew) Full of grace.
Nancy (Hebrew) Full of grace.
Nanine (Hebrew) Full of grace.
Nannette (Hebrew) Full of grace.
Nannie (Hebrew) Full of grace.
Nathalia (Latin) Natal day.
Natividad (Latin) Natal day.
Neille (Gaelic) The champion.
Nellie (Old French) Light.
Neoma (Greek) New moon.
Netty (Hebrew) Full of grace.
Nichole (Greek) The people's victory.
Ninon (Spanish) Girl.
Nisse (Norse) Friendly elf or brownie.
Nissie (Norse) Friendly elf or brownie.
Nolana (Latin) Symbol of peace.
Nonie (Latin) The ninth.
Nora (Greek) Light.
Norine (Latin) A rule, pattern, or precept.
Oceana (Greek) Ocean.
Odell (Hebrew) I will praise God.
Ofelia (Greek) Serpent.

Olexa (Greek) Helper of humankind.
Ona (Latin) Unity.
Ondrea (Latin) Womanly.
Oneida (American Indian) The looked-for one.
Ophelia (Greek) Serpent.
Orenda (American Indian) Magical power.
Pam (Greek) All honey.
Pamela (Greek) All honey.
Pansy (Greek) Flowerlike.
Pascale (Latin) Born at Easter or Passover.
Patti (Latin) Wellborn, noble.
Peace (Old English) The peaceful one.
Peony (Greek) Praise-giving.
Petula (Latin) Seeker.
Philomena (Greek) Loving song.
Pier (Latin) Rock or stone.
Posy (English) Small bunch of flowers.
Prima (Latin) First or first child.
Prisca (Latin) From ancient times.
Prudy (Latin) Foresight.
Querida (Spanish) Beloved.
Rabiah (Arabic) Breeze.
Rachael (Hebrew) Innocent as a lamb.
Raisa (Greek) A rose.
Ramonda (Old German) Mighty or wise protector.

Raynice (Sanskrit) Queen.
Reanne (Welsh) A witch.
Renell (Latin) Reborn.
Rita (Latin) A pearl.
Rohana (Hindi) Sandalwood.
Rona (Old English) Mighty and powerful.
Rosalia (Latin) Rose.
Rosalinda (Spanish) Beautiful rose.
Rosaline (Spanish) Beautiful rose.
Rose (Greek) A rose.
Rosette (Greek) A rose.
Rosie (Greek) A rose.
Rui (Japanese) Troublesome.
Russet (English) A reddish or yellowish-brown color.
Sachiko (Japanese) Bliss child, joy.
Sahara (Arabic) Wilderness.
Sandra (Greek) Helper of humankind.
Sara (Hebrew) Of royal status.
Saree (Arabic) Most noble.
Savina (Latin) Sabine woman.
Shanley (Gaelic) Child of the old hero.
Sharik (African) God's child.
Sharon (Hebrew) A level plain.
Sharron (Hebrew) A level plain.
Shawna (Hebrew) God's gracious gift.
Sherry (Old French) Beloved.
Shulamith (Hebrew) Peace.

Sibylle (Greek) Prophetess.
Sigrid (Old German) Victorious protector.
Simone (Hebrew) One who hears.
Simonette (Hebrew) One who hears.
Sirena (Greek) A siren.
Solenne (Spanish) Sunshine.
Starlin (English) Star.
Sugar (English) Sugar.
Sunny (English) Bright, cheerful.
Susanetta (Hebrew) Graceful lily.
Susanne (Hebrew) Graceful lily.
Tangerine (English) Tangerine-colored.
Tansey (Greek) Immortality.
Tatiana (Russian) Fairy queen.
Tatum (Old English) Cheerful.
Tereza (Greek) The harvester.
Tia (Spanish) Aunt.
Tine (Greek) Anointed.
Tonya (Latin) Praiseworthy, without a peer.

Trisha (Latin) Wellborn, noble.
Trudey (Old German) Beloved.
Tullia (Gaelic) Peaceful one.
Ulani (Polynesian) Cheerful.
Verita (Latin) Truth.
Verona (Latin) Lady of Verona.
Vilette (Old French) From the village.
Vilma (Old German) Determined guardian.
Willa (Old German) Determined guardian.
Winonah (American Indian) Firstborn daughter.
Yasu (Japanese) The tranquil.
Yedda (Old English) The singer.
Yoko (Japanese) The positive female.
Yoshiko (Japanese) Good.
Zelda (German) Grey battle-maiden.
Zivah (Israeli) Radiant.

NUMBER THREE NAMES—MALE

Addison (Old English) Son of Adam.

Alban (Latin) White.

Alec (Greek) Helper of humankind.

Alexander (Greek) Helper of humankind.

Alexandre (Greek) Helper of humankind.

Alexio (Greek) Helper of humankind.

Alister (Greek) Helper of humankind.

Allard (Old English) Noble and brave.

Amadeo (Latin) Love of God.

Amos (Hebrew) Burden.

Argus (Greek) Watchful.

Armando (Old German) Army man.

Army (Old German) Army man.

Arnan (Hebrew) Quick, joyful.

Arno (Old German) Strong as an eagle.

Art (Celtic-Welsh) Noble, bear-hero.

Asa (Hebrew) Physician.

Augustus (Latin) Majestic dignity.

Balfour (Scottish Gaelic) Pasture land.

Barrett (Old German) Mighty as a bear.

Beck (Old English) A brook.

Ben (Hebrew) Son of my right hand.

Bern (Old German) Brave as a bear.

Berthold (Old German) Brilliant ruler.

Blase (Latin) Stammerer.

Bradleigh (Old English) From the broad meadow.

Brand (Old English) Firebrand.

Breck (Gaelic) Freckled.

Brien (Celtic) Strength with virtue and honor.

Burke (Old French) Dweller at the fortress.

Cale (Hebrew) Spontaneous.

Carlin (Gaelic) Little champion.

Carney (Gaelic) Victorious.

Charles (Old German) Strong and masculine.

Chen (Chinese) Great.

Colbert (Old English) Brilliant seafarer.

Colby (Old English) From the dark farm.

Connell (Celtic) Mighty, valorous.

Curran (Gaelic) Champion or hero.

Dag (Scandinavian) Day, brightness.

Dalton (Old English) From the estate in the valley.

Darnell (Old English) From the hidden place.

Davey (Hebrew) Beloved.

Delaney (Gaelic) Descendant of the challenger.

Delbert (Old English) Bright as day.

Delmer (Old French) From the sea.

Delvin (Old English) Godly friend.

Denzel (English) From Denzel in Cornwall.

Dermot (Gaelic) Free from envy.

Deryck (Old German) Leader of the people.

Devlin (Gaelic) Fierce valor.

Diccon (Old German) Wealthy, powerful.

Dillon (Gaelic) Faithful.

Drake (Old English) Dragon.

Duncan (Scottish Gaelic) Dark-skinned warrior.

Dunstan (Old English) From the brown stone hill.

Eamon (Old English) Prosperous guardian.

Edison (Old English) Son of Edward.

Edric (Old English) Power and good fortune.

Eldred (Old English) Sage counsel.

Elliott (Hebrew) Jehovah is God.

Ellis (Hebrew) Jehovah is God.

Eloy (Latin) Chosen one.

Elroy (Old French) The King.

Elsworth (Old English) Nobleman's estate.

Elton (Old English) From the old town.

Emery (Old German) Industrious ruler.

Emil (Latin) Flattering.

Emmanuel (Hebrew) God is with us.

Esteban (Greek) Crowned.

Ethan (Hebrew) Firm or strong.

Eugene (Greek) Wellborn noble.

Fabiano (Latin) Bean grower.

Fairleigh (Old English) From the sheep meadow.

Farleigh (Old English) From the sheep meadow.

Ferdinand (Old German) World-daring.

Ferris (Old English) Iron-strong.

Fielding (Old English) From the field.

Findley (Gaelic) Little fair-haired soldier.

Fleming (Old English) Dutchman.

Florian (Latin) Flowering or blooming.

Fontaine (Old French) Fountain, spring.

Galen (Greek) Calm.

Geoff (Old German) God's divine peace.

George (Greek) Landholder, farmer.

Gibson (Old English) Son of Gilbert.

Gitano (Old English) Wanderer.

Graham (Old English) The grey home.

Gram (Old English) The grey home.

Grantley (Old English) From the great plains.

Gunnar (German) Battler, warrior.

Gunther (German) Battler, warrior.

Gwayne (Celtic) The battle hawk.

Hal (Old German) Lord of the manor.

Hamlin (Old French) The home lover.

Harper (Old English) Harp player.

Harrison (Old English) Son of Harry.

Hayden (Old English) From the hedged valley.

Henderson (Old English) Son of Henry.

Herald (Old English) Army commander.

Hershel (Hebrew) Deer.

Holton (Old English) From the town in the hollow.

Hosea (Hebrew) Salvation.

Howell (Celtic) Little, alert.

Humberto (Old German) Little Hun.

Izaak (Hebrew) He laughs.

James (Hebrew) The supplanter.

Jarman (Old German) The German.

Jayson (Greek) Healer.

Jeffery (Old German) God's divine peace.

Jeffrey (Old German) God's divine peace.

Jermaine (Latin) A German.

Jerome (Latin) Sacred or holy name.

Jock (Hebrew) The supplanter.

Joe (Hebrew) He shall add.

Jon (Hebrew) God's gracious gift.

Jonah (Hebrew) Dove.

Judd (Hebrew) Praised.

Juel (Greek) Youthful, downy-bearded.

Justin (Latin) Just.

Kade (Scottish Gaelic) From the wetlands.

Kamden (Scottish Gaelic) From the winding valley.

Kaspar (Persian) Treasurer.

Keaton (Old English) Place of the hawks.

Ken (Gaelic) Handsome.

Kendrick (Old English) Royal ruler.

Kentaro (Japanese) Big boy.

Kenyon (Gaelic) White-haired blonde.

Kerrick (Old English) King's rule.

Keven (Gaelic) Gentle, kind, and lovable.

Kingsley (Old English) From the king's meadow.

Kord (Old French) Ropemaker.

Korrigan (Celtic) Spearman.

Lamont (Old Norse) Lawyer.

Langston (Old English) From the long, narrow town.

Leland (Old English) Meadow land.

Leonardo (Old French) Lion-brave.

Leroy (Old French) King.

Lev (Latin) Lion.

Levi (Hebrew) Joined, united.

Linus (Greek) Flaxen-haired.

Llewellyn (Welsh) Lionlike or lightning.

Loring (Old German) Son of the famous warrior.

Luther (Old German) Famous warrior.

Lyndon (Old English) From the linden tree hill.

Magnus (Latin) Great.

Major (Latin) Greater.

Makoto (Japanese) Good.

Manuel (Hebrew) God is with us.

Marcus (Latin) Follower of Mars.

Marshall (Old French) Steward, horse-keeper.

Martin (Latin) Follower of Mars.

McDonald (Celtic) Son of Donald.

Melvin (Gaelic) Brilliant chief.

Meyer (Hebrew) Bringer of light.

Michio (Japanese) Man with strength of three thousand.

Monolo (Hebrew) God is with us.

Moore (Old French) Dark-complected.

Morell (Old French) The Moor.

Mort (Old French) Still water.

Mortimer (Old French) Still water.

Moss (Hebrew) Drawn out of the water.
Nahum (Hebrew) Consoling.
Natal (Latin) Natal day.
Nathaniel (Hebrew) Gift of God.
Neron (Latin) Stern.
Newbold (Old English) From the new building.
Niall (Gaelic) The champion.
Noble (Latin) Wellborn.
Norman (Old French) A northman.
Norris (Old French) Northerner.
North (Old English) From the north farm.
Norward (Old English) Guardian of the north.
Nyles (Gaelic) The champion.
Octave (Latin) The eighth.
Oran (Gaelic) Pale-complected.
Orville (Old French) From the golden estate.
Osgood (Old English) Divinely good.
Owen (Old Welsh) Wellborn young warrior.
Pancho (Latin) Free.
Parnel (Old French) Little Peter.
Paton (Latin) Wellborn, noble.
Pearce (Latin) Rock or stone.
Per (Latin) Rock or stone.
Radford (Old English) From the reedy ford.
Rafe (Hebrew) Healed by God.

Rawley (Old English) From the deer meadow.
Rey (Latin) King.
Reynold (Old English) Mighty and powerful.
Ricco (Old German) Wealthy, powerful.
Richmond (Old German) Powerful protector.
Ring (Old English) Ring.
Roan (Old English) From the rowan tree.
Roberto (Old English) Bright fame.
Rochester (Old English) From the stone camp.
Roddy (Old English) One who rides with a knight.
Romeo (Italian) Pilgrim to Rome.
Roscoe (Old Norse) From the deer forest.
Royce (Old English) Son of the king.
Rush (Old French) Red-haired.
Ryleigh (Gaelic) Valiant.
Sargent (Old French) Army officer.
Schuyler (Dutch) Sheltering.
Scot (Old English) Scotsman.
Severo (Latin) Strict, restrained.
Shep (Old English) Shepherd.
Silvester (Latin) From the forest.
Skelly (Gaelic) Storyteller.

Stanwood (Old English) From the rocky wood.

Steele (Old English) Hard, durable.

Sylvan (Latin) From the forest.

Tayib (Indian) Good or delicate.

Templeton (Old English) From the town of the temple.

Terance (Latin) Smooth.

Terrance (Latin) Smooth.

Thaine (Old English) Attendant warrior.

Thane (Old English) Attendant warrior.

Theo (Greek) Gift of God.

Thorn (Old English) From the thorny embankment.

Tobias (Hebrew) The Lord is good.

Tod (Old English) A fox.

Toddie (Old English) A fox.

Tom (Hebrew) The devoted brother.

Trahern (Welsh) Strong as iron.

Traver (Old French) At the crossroads.

Trevar (Gaelic) Wise and discreet.

Tripper (Old English) To dance or hop.

Tyson (Old French) Firebrand.

Ulysses (Latin-Greek) Wrathful.

Uriah (Hebrew) Jehovah is my light.

Valentine (Latin) Strong.

Victorio (Latin) Victorious.

Vidal (Latin) Life.

Vinny (Latin) Victorious.

Vinson (Old English) The conqueror's son.

Vito (Latin) Alive.

Wardell (Old English) Guardian.

Waverley (Old English) Quaking aspen tree meadow.

Wendell (Old German) The wanderer.

Willis (Old German) Determined guardian.

Wilton (Old English) From the town by the spring.

Wood (Old English) From the passage in the woods.

Worth (Old English) From the farmstead.

Xenophon (Greek) Strange voice.

Zachariah (Hebrew) Jehovah has remembered.

NUMBER THREE NAMES—UNISEX

Abby (Hebrew) My father is joy.

Adriaan (Latin) Black earth.

Allie (Greek) Truthful.

Angel (Greek) Heavenly messenger.

Ardelle (Latin) Ardent, fiery.

Austin (Latin) Majestic dignity.

Blaise (Latin) Stammerer.

Bobbi (Old English) Bright fame.

Brandi (Dutch) Brandy drink.

Brooke (Old English) From the brook.

Cace (Gaelic) Vigilant.

Chris (Greek) Anointed.

Clare (Latin) Bright, shining girl.

Corey (Gaelic) From the hollow.

Fran (Latin) Free.

Franky (Latin) Free.

Garland (Old French) Wreath.

Georgie (Greek) Landholder, farmer.

Geri (Old German) Spear-mighty.

Hayatt (Old English) From the high gate.

Hollis (Old English) From the grove of holly trees.

Jackie (Hebrew) The supplanter.

Jean (Hebrew) God's gift of grace.

Julie (Greek) Young in heart and mind.

Kai (Old Welsh) Keeper of the keys.

Karol (Old German) Noble and strong.

Kodi (Old English) A cushion.

Kris (Greek) Anointed.

Lanny (Old English) From the narrow road.

Laurie (Latin) Crowned with laurels.

Lindsay (Old English) From the linden tree island.

Linsey (Old English) From the linden tree island.

Lisle (Old French) From the island.

Lou (Old German) Famous warrior.

Lyell (Old French) From the island.

Madison (Old English) Son of the powerful soldier.

Malory (Old German) Army counselor.

Mel (Gaelic) Brilliant chief.

Nealy (Gaelic) The champion.

Nike (Greek) The people's victory.

Quinn (Gaelic) Wise, intelligent.

Rabi (Arabic) Breeze.

Ricky (Old German) Wealthy, powerful.

Ronnie (Old English) Mighty and powerful.

Rowen (Gaelic) Red-haired.

Ruby (Old French) Ruby gemstone.

Sasha (Greek) Helper of humankind.

Scotty (Old English) Scotsman.

Sean (Hebrew) God's gracious gift.

Simeon (Hebrew) One who hears.

Sloane (Gaelic) Warrior.

Stace (Latin) Stable, prosperous.

Stacie (Latin) Stable, prosperous.

Tailor (Old English) Tailor.

Tommie (Hebrew) The devoted brother.

Wallace (Old English) From Wales.

Whitley (Old English) From the white field.

8

Number Four/Twenty-Two

Qualities of Number Four

Keywords: practical, methodical, solid, studious, punctual, organized, structured, discerning, visionary, useful, orderly, conscientious, industrious, painstaking, thorough, systematic, steadfast, reliable, conservative, patient, loyal, serious, narrow-minded, spiteful, stubborn, slow, indifferent, overcautious, stingy, reactionary, rigid, stern, strict, dogmatic, inhibited, dull.

Symbolism: The number four represents foundations. Four is feminine in nature and stands for the constructive builder. Its symbols are Uranus and the square, and its element is air. Four has always been a sacred number, and to the Pythagoreans it denoted completion, solidity, stability, and equilibrium. It is related to the earth, the establishment, solid matter, the points of the compass, the seasons, the elements, and the winds.

Personality: The number four personality is a builder and organizer. Fours know how to get things together and keep them functioning properly. They have strength of purpose and willpower, and they are efficient, industrious, and extremely practical. Fours are the quintessential middle managers, loyal to employers and generally fair to employees.

Since four is the number of physical manifestation, they like to see the results of their work. There is a deep need for stability, which causes them to build strong houses and institutions. They take the ideas of other people and put them to practical use. Fours are tenacious, precise, and calm, and they tend to deal with life carefully and systematically. Down to earth, solid, and respectable, Fours find their happiness and fulfillment in regular endeavor. They do not like change. Fours love their homes and are usually faithful and devoted to their mates. They are excellent money managers and rarely live beyond their means. Fours are generally trustworthy and can be relied on to keep their word.

Fours are often viewed as dull or boring because they spend so little time having fun. When overworked, Fours may suffer severe bouts of depression. They can easily get stuck in a negative mold and become repressed and/or repressive. When Fours refuse to take responsibility for their own actions, they look for someone else to blame when things go wrong.

Careers: accountant, architect, astrologer, aviator, banker, bookkeeper, builder, chemist, computer expert, electrician, engineer, inventor, investigator, machinist, manufacturer, mathematician, mechanic, occultist, plumber, printer, radiologist, reformer, scientist, sculptor, teacher.

Qualities of number Twenty-Two (A master number and the higher form of four)

Keywords: master builder, creative, wise, courageous, forceful, hardworking, successful, masterful, spiritual, understanding, intelligent, magnificent, accomplished, enlightened, practical, idealistic, progressive, selfless, artistic, fanatical, extremist, extravagant, criminal, willful, contrary, bigoted, evil.

Symbolism: The number twenty-two is a master number, and as the higher form of four it represents the master builder. Twenty-two stands for dominance of the material and spiritual planes. Those with the number twenty-two as a major vibration are said to have mastered the physical and in so doing have it to command. This gives them the power to build structures for the betterment of humankind. Twenty-two is the number of leadership, and those with this vibration are said to come into society in order to help shoulder its physical responsibilities.

Personality: The number twenty-two personality is magnetic and powerful and interested in everything new or revolutionary. Often said to have been born ahead of their time, Twenty-twos are the luminaries who light the way to the future. As masters of both planes (physical and spiritual), they have the ability to make inspirational things practical. Anything they wish to accomplish in a material way is possible for them. Their use of higher laws to control and manipulate their own environment enables them to give wise counsel to others. They are courageous, forceful, hardworking, and are often described as workaholics. Some have very marked leanings toward mysticism and the occult. No door is closed to them, and anything that they wish to accomplish in a material way is possible. As planners, organizers, and doers, Twenty-twos can be builders of beautiful structures. Some become famous as actors, statesmen, or explorers. Whatever they can visualize is theirs to build, and many Twenty-twos achieve greatness during their lifetimes.

Twenty-twos can turn their fine brains toward criminal activities and become the masterminds behind daring crimes. Sometimes the power to do great things for humanity turns into a quest for personal power and a need to control every situation. When this happens the Twenty-two's vibration may tear down and destroy rather than build. Whenever they lose sight of their potential for great deeds and selfless accomplishments, they risk becoming merely greedy and ruthless.

Careers: accountant, architect, actor, astrologer, bookkeeper, diplomat, executive, fund-raiser, lawyer, musician, philanthropist, statesman.

NUMBERS FOUR AND TWENTY-TWO NAMES—FEMALE

Abra (Hebrew) Earth mother.
Adalicia (Old German) Noble.
Adel (Old German) Of noble rank.
Ainsley (Scottish Gaelic) From one's own meadow.
Alarice (Old German) Rules all.
Alcina (22/4) (Greek) Strong-minded.
Aliza (22/4) (Greek) Truthful.
Amantha (22/4) (Greek) Beyond death.
Aminta (22/4) (Greek) Protector.
Anamaria (Hebrew) Bitter, graceful.
Anastasia (22/4) (Greek) Of the resurrection.
Angela (22/4) (Greek) Heavenly messenger.
Angeline (Greek) Heavenly messenger.
Annabel (22/4) (Hebrew-Latin) Graceful, beautiful.
Anora (22/4) (Old English) Light and graceful.
Anthea (22/4) (Greek) Like a flower.
Araceli (Latin) Altar of heaven.
Arsenia (Greek) Potent.

Athena (22/4) (Greek) Wisdom.
Aurelia (Latin) Golden.
Averil (Latin) Forthcoming.
Azura (22/4) (Old French) Sky blue.
Bernadette (Old German) Brave as a bear.
Beulah (22/4) (Hebrew) Married.
Beverley (Old English) From the beaver meadow.
Bibi (22/4) (Latin) Alive.
Blyth (22/4) (Old English) Joyful.
Brena (22/4) (Gaelic) Little raven.
Brittania (Latin) From England.
Brunella (Old German) Brown-haired.
Bunny (22/4) (English) Little rabbit.
Burgundy (French) A reddish-purple color.
Calvinna (Latin) Bald.
Candace (22/4) (Greek) Glittering, glowing white.
Candide (Greek) Glittering, glowing white.
Cari (22/4) (Turkish) Flows like water.

Cedra (22/4) (Old English) Battle chieftain.

Chandra (Sanskrit) Moonlike.

Chantalle (Latin) Song.

Charis (Greek) Grace.

Clarabella (French) Brilliant, beautiful.

Clorinda (Latin) Famed for her beauty.

Cloris (Greek) Pale.

Clotilda (old German) Famous battle maiden.

Columba (22/4) (Latin) Dovelike, gentle.

Columbia (Latin) Dovelike, gentle.

Constance (Latin) Firmness, constancy.

Cookie (English) Cookie.

Coral (22/4) (Latin) Small sea creature.

Cordelia (Welsh) Jewel of the sea.

Cozette (French) Little pet.

Cristobel (Greek) Beautiful anointed one.

Cynthera (Greek) Goddess of love and beauty.

Daisy (22/4) (Old English) Eye-of-the-day.

Danika (22/4) (Slavic) The morning star.

Delia (22/4) (Greek) Visible.

Demi (22/4) (Latin) Half, partly belonging.

Denice (Greek) Wine goddess.

Devi (22/4) (Sanskrit) Divine.

Devony (Old English) Defender of Devonshire.

Diza (22/4) (Hebrew) Joy.

Donelle (Latin) Lady.

Dyane (22/4) (Latin) Divine moon goddess.

Echo (22/4) (Greek) Sound.

Edythe (Old English) Rich gift.

Effie (Greek) Well spoken of.

Elma (Old German) God's protection.

Elspeth (Hebrew) God's promise.

Elva (Old English) Elfin, good.

Elvia (22/4) (Old English) Elfin, good.

Elvira (Latin) White, blond.

Emanuele (Hebrew) God is with us.

Emerald (French) Emerald jewel.

Emilia (Latin) Flattering.

Emmeline (Old German) One who leads the universe.

Eriko (Japanese) Child with a collar.

Evangeline (Greek) Bearer of good tidings.

Fannie (Latin) Free.

Fatimah (Arabic) Wean or abstain.

Fawne (22/4) (Old French) Young deer.

Fedora (Greek) Gift of God.
Felice (Latin) Joyous, fortunate.
Felicie (Latin) Joyous, fortunate.
Felipa (Greek) Lover of horses.
Fionn (Gaelic) Fair.
Flossie (Latin) Flowering or blooming.
Frannie (Latin) Free.
Garda (22/4) (Old German) The protected.
Genevieve (French) Pure white wave.
Georgina (Greek) Landholder, farmer.
Gina (22/4) (Latin) Queen.
Gipsy (Old English) Wanderer.
Godiva (Old English) Gift of God.
Gretel (Latin) A pearl.
Gwen (22/4) (Old Welsh) White-browed.
Halia (22/4) (Hawaiian) Remembrance of a beloved.
Harmony (Latin) Harmony.
Hedda (22/4) (Old German) Strife.
Helise (Old German) Famous woman warrior.
Hildy (Old German) Battle maid.
Honey (Old English) Sweet.
Iduna (22/4) (Old Norse) Lover.
Illana (Hebrew) Tree.
Illiana (Greek) Light, a torch.

Inga (22/4) (Norse) Hero's daughter.
Isabeau (22/4) (Old Spanish) Consecrated to God.
Isadora (Greek) Gift of Isis.
Jacinta (22/4) (Greek) Hyacinth flower.
Jamilla (22/4) (Arabic) Beautiful.
Janeen (22/4) (Hebrew) God's gift of grace.
Janna (Arabic) A harvest of fruit.
Jeanne (22/4) (Hebrew) God's gift of grace.
Jeannette (Hebrew) God's gift of grace.
Jenifer (French) Pure white wave.
Jillian (Greek) Young in heart and mind.
Josette (22/4) (Hebrew) He shall add.
Josie (22/4) (Hebrew) He shall add.
Joyce (22/4) (Latin) Joyous.
Juanita (22/4) (Hebrew) God's gracious gift.
Justina (22/4) (Latin) Just.
Kaitlin (Greek) Pure maiden.
Kalifa (22/4) (Somali) Chaste, holy.
Kallista (22/4) (Greek) Most beautiful woman.
Kara (Greek) Pure maiden.

Karen (22/4) (Greek) Pure maiden.

Karis (22/4) (Greek) Grace.

Karmina (Latin) Song.

Kathleen (Greek) Pure maiden.

Keely (22/4) (Gaelic) Beautiful.

Kiara (22/4) (Greek) Pure maiden.

Kitty (22/4) (Greek) Pure maiden.

Korin (Greek) The maiden.

Kyla (Gaelic) Handsome.

Leala (French) Loyal, faithful.

Leda (Latin) Gladness.

Leeza (22/4) (Hebrew) God's promise.

Leontine (Latin) Lionlike.

Letitia (Latin) Gladness.

Lia (Hebrew) Weary.

Lily (22/4) (Latin) Lily flower.

Linda (22/4) (Spanish) Pretty.

Linette (Old French) Songbird.

Lisabeth (Hebrew) God's promise.

Lisha (22/4) (Arabic) The darkness before midnight.

Lola (Spanish) Sorrows.

Lorelei (German) Siren.

Lourdes (Basque) Rocky point.

Luana (Old German-Hebrew) Graceful battle maid.

Ludella (22/4) (Old English) Pixie maid.

Lyric (Latin) Of the lyre.

Lysandra (Greek) The liberator.

Magdalane (Hebrew) Woman of Magdala.

Magdalena (Hebrew) Woman of Magdala.

Malva (Greek) Soft and tender.

Manuela (22/4) (Hebrew) God is with us.

Marcie (Latin) Follower of Mars.

Mariel (Hebrew) Bitterness.

Mariko (Japanese) Ball, circle.

Marney (Latin) From the sea.

Martina (Latin) Follower of Mars.

Maya (Sanskrit) Illusion.

Megan (22/4) (Gaelic) Soft, gentle.

Melinda (Greek) Dark-clothed.

Melly (22/4) (Greek) Dark-clothed.

Melvina (Gaelic) Brilliant chief.

Mercia (Latin) Compassion.

Michelle (Hebrew) Like unto the Lord.

Milagros (Latin) Miracles.

Miliani (Hawaiian) Gentle caress.

Missy (22/4) (Greek) Honeybee.

Monique (Latin) Advisor.

Morwenna (Welsh) Ocean waves.

Nastassia (22/4) (Latin) Natal day.

Natala (Latin) Natal day.

Natalee (22/4) (Latin) Natal day.

Natalia (22/4) (Latin) Natal day.

Neona (22/4) (Greek) New moon.

Nerissa (Greek) Of the sea.

Nerita (Greek) Of the sea.

Nicole (Greek) The people's victory.

Nicolette (Greek) The people's victory.

Noleta (22/4) (Latin) Unwilling.

Norell (Scandinavian) From the north.

Odella (22/4) (Hebrew) I will praise God.

Ophra (Hebrew) Young deer.

Oprah (Hebrew) Young deer.

Oriana (Latin) Dawning, golden.

Paloma (22/4) (Spanish) Dove.

Patci (22/4) (Latin) Wellborn, noble.

Pepita (Hebrew) He shall add.

Perrine (Latin) Rock or stone.

Persephone (Greek) Goddess of the underworld.

Phillis (Greek) A green branch.

Pippa (Greek) Lover of horses.

Princesa (Latin) Princess.

Princess (Latin) Princess.

Psyche (Greek) Soul or mind.

Queenie (Old English) Queen.

Raeanne (Hebrew) Innocent as a lamb.

Reeta (22/4) (Latin) A pearl.

Reiko (Japanese) Gratitude, propriety.

Riannon (Welsh) A witch.

Rica (22/4) (Old German) Peaceful ruler.

Roana (22/4) (Greek-Hebrew) A rose, full of grace.

Roanne (Greek-Hebrew) A rose, full of grace.

Romona (Old German) Mighty or wise protector.

Rosabella (French-Greek) Beautiful rose.

Rosemarie (English) Rose of Saint Mary.

Rosene (Greek) A rose.

Roshelle (French) From the little rock.

Rosina (Greek) A rose.

Roslyn (Spanish) Beautiful rose.

Rowena (Old English) Well-known friend.

Royale (Old French) Royal.

Ruth (22/4) (Hebrew) Compassionate.

Sachi (22/4) (Japanese) Bliss child, joy.

Sasa (Japanese) Help, aid.

Seana (Hebrew) God's gracious gift.

Secunda (22/4) (Latin) Second-born.

Sephira (Hebrew) Ardent.

Sharona (Hebrew) A level plain.

Sharyn (Hebrew) A level plain.

Shirlee (Old English) From the bright meadow.

Shoshana (Hebrew) Graceful lily.

Shulamit (Hebrew) Peace.

Sibyl (22/4) (Greek) Prophetess.

Silver (Old English) Silver-colored metal.

Sonia (22/4) (Greek) Wisdom.

Star (English) Star.

Starlene (English) Star.

Starr (22/4) (English) Star.

Sumer (22/4) (Old English) Summer.

Sybil (22/4) (Greek) Prophetess.

Tacy (Latin) To be silent.

Taffy (22/4) (Welsh) Beloved.

Tamsin (22/4) (Hebrew) The devoted sister.

Tangy (22/4) (English) Tangerine-colored.

Tara (Gaelic) Rocky pinnacle.

Taree (22/4) (Japanese) Bending branch.

Tasha (Latin) Natal day.

Theresa (Greek) The harvester.

Tiara (22/4) (Latin) Headdress.

Tiena (22/4) (Greek) Anointed.

Tillie (Old German) Powerful in battle.

Toni (22/4) (Latin) Praiseworthy, without a peer.

Torie (Latin) Victorious.

Torrie (Latin) Victorious.

Tova (Hebrew) God is good.

Trixie (Latin) Bringer of joy.

Valda (Old German) The ruler.

Valencia (Latin) Strong.

Venetia (Latin) From Venice.

Verla (22/4) (Latin) Truth.

Vernice (Latin) Springlike, youthful.

Vesta (Latin) She who dwells.

Vitoria (Latin) Victorious.

Wallis (22/4) (Old English) From Wales.

Wendye (Old German) The wanderer.

Wilda (22/4) (Old English) Willow.

Willow (Old English) Willow.

Wilma (22/4) (Old German) Determined guardian.

Winona (American Indian) Firstborn daughter.

Yoshi (Japanese) Good.

Yumiko (Japanese) Lily child.

Zerlina (Old German) Serene beauty.

Zitella (Latin) A little rose.

NUMBERS FOUR AND TWENTY-TWO NAMES—MALE

Aaron (22/4) (Hebrew) Lofty or exalted.

Abbot (Hebrew) My father is joy.

Abdul (Arabic) Servant of God.

Abner (22/4) (Hebrew) Father of light.

Adiel (22/4) (Hebrew) God is an ornament.

Adler (22/4) (Old German) Eagle.

Adney (22/4) (Old English) Dweller on the island.

Adriel (Hebrew) Of God's flock.

Ahmed (22/4) (Arabic) Praiseworthy.

Aiken (22/4) (Old English) Little Adam.

Albert (22/4) (Old English) Noble and brilliant.

Alby (Latin) White.

Alexandro (Greek) Helper of humankind.

Allain (22/4) (Gaelic) Handsome, cheerful.

Allan (Gaelic) Handsome, cheerful.

Alonso (22/4) (Old German) Noble and eager.

Aloysius (Old German) Famous in war.

Alvin (22/4) (Old German) Beloved by all.

Amiel (22/4) (Hebrew) Lord of my people.

Annell (22/4) (Celtic) Beloved.

Ansley (22/4) (Old English) From Ann's meadow.

Archibald (Old German) Genuinely bold.

Aric (22/4) (Old English) Sacred ruler.

Arnall (22/4) (Old German) Strong as an eagle.

Auberon (Old German) Elf ruler.

Aubert (22/4) (Old English) Noble and brilliant.

Audie (22/4) (Old English) Old friend.

Avenall (22/4) (Old French) Dweller in the oat field.

Baldric (Old German) Princely ruler.

Barnabas (22/4) (Greek) Son of prophecy.

Basir (22/4) (Turkish) Intelligent and discerning.

Berlyn (Old English) Son of Burl.

Birch (Old English) Birch tree.

Boden (22/4) (Old English) Herald, messenger.

Borden (Old English) From the valley of the boar.

Braddock (Old English) Broad-spreading oak.

Bradley (Old English) From the broad meadow.

Bradshaw (Old English) From the large virginal forest.

Braxton (Old English) From Braxton.

Brendan (Gaelic) Little raven.

Brigham (Old English) From the enclosed bridge.

Brock (22/4) (Old English) Badger.

Broderick (Old English) From the broad ridge.

Bruce (22/4) (French) From the thicket.

Budd (Old English) Herald, messenger.

Burney (Old English) Brook.

Camden (22/4) (Scottish Gaelic) From the winding valley.

Cantrell (Latin) Singer.

Carlo (22/4) (Old German) Strong and masculine.

Carlyle (Old English) From the fortified town.

Carmelo (Latin) Song.

Carr (22/4) (Old Norse) He who dwells by the marsh.

Carver (Old English) Woodcarver.

Cash (Latin) Vain.

Caspar (22/4) (Persian) Treasurer.

Castor (22/4) (Greek) Industrious.

Christopher (Greek) Bearer of Christ.

Churchill (Old English) By the church on the hill.

Clancy (22/4) (Celtic) Redheaded.

Cleon (22/4) (Greek) Famous.

Cliffton (Old English) From the cliff-town.

Clint (22/4) (Old English) From the hillside.

Clyde (22/4) (Celtic) Heard from afar.

Cornel (Latin) Yellow, horn-colored.

Corrigan (Celtic) Spearman.

Cullen (22/4) (Gaelic) Handsome.

Cyril (Greek) Master, ruler.

Dalston (22/4) (Old English) From Daegal's town.

Darrel (Old French) Beloved.

David (22/4) (Hebrew) Beloved.

Delwin (Old English) Proud friend.

Dexter (Latin) Right-handed.

Diego (Hebrew) The supplanter.
Dietrich (Old German) Ruler of the people.
Dionisio (Greek) God of wine.
Dominic (Latin) Belonging to the Lord.
Donnell (Gaelic) World ruler.
Donovan (Gaelic) Dark warrior.
Dugald (22/4) (Celtic) From the dark stream.
Edwald (22/4) (Old English) Prosperous ruler.
Efron (Hebrew) Young stag.
Elden (22/4) (Old English) Wise protector.
Elston (22/4) (Old English) Nobleman's town.
Elvern (Latin) Spring, greening.
Elvis (22/4) (Norse) The prince of wisdom.
Emerys (Greek) Immortal.
Emmett (22/4) (Old German) Industrious ruler.
Emory (Old German) Industrious ruler.
Erastus (22/4) (Greek) Beloved.
Erle (22/4) (Old English) Nobleman.
Ewing (Old English) Friend of the law.
Falkner (Old English) Trainer of falcons.
Farley (Old English) From the sheep meadow.
Fergus (Gaelic) The best choice.

Ferron (Old English) Ironworker.
Finlay (Gaelic) Little fair-haired soldier.
Fitzpatrick (Old English) Son of a nobleman.
Fonda (22/4) (Latin) The profound.
Fortunato (Latin) Fortunate.
Franklin (Latin) Free.
Fraser (Old English) Curly-haired.
Fulbright (Old German) Very bright.
Gaelan (22/4) (Gaelic) Intelligent.
Gardner (Old English) A gardener.
Garrad (Old English) Spear-strong.
Garrard (Old English) Spear-strong.
Garrick (Old English) Oak spear.
Gaston (22/4) (French) From Gascony.
Gaven (22/4) (Old Welsh) White hawk.
German (Old German) People of the spear.
Gershom (Hebrew) Exile.
Gian (22/4) (Hebrew) God's gracious gift.
Gill (22/4) (Old German) Brilliant hostage.

Glyn (22/4) (Old Welsh) Dweller in a valley.

Gordie (Old English) Round hill.

Gregorio (Latin) Watchman.

Grover (Old English) From the grove.

Guillermo (Old German) Determined guardian.

Gustavus (22/4) (Swedish) Staff of the Goths.

Hamish (Hebrew) The supplanter.

Harold (Old German) Lord of the manor.

Hashim (Arabic) Destroyer of evil.

Haslett (22/4) (Old English) From the hazel tree land.

Hayes (22/4) (Old English) From the hedged place.

Herbert (Old German) Glorious soldier.

Hersh (Hebrew) Deer.

Hillel (Hebrew) Greatly praised.

Hiram (Hebrew) Most noble.

Holden (Old English) From the hollow in the valley.

Houston (Old English) From the town on the hill.

Howey (Old English) Cliff guardian.

Ignacio (Latin) Fiery or ardent.

Igor (Norse) Famous son.

Ingemar (Norse) Famous son.

Ishmael (Hebrew) God will hear.

Jacob (Hebrew) The supplanter.

Jacques (22/4) (Hebrew) The supplanter.

Jaedon (22/4) (Hebrew) Jehovah has heard.

Jeremy (Hebrew) Exalted by the Lord.

Jethro (Hebrew) Preeminence.

Jose (Hebrew) He shall add.

Jotham (22/4) (Hebrew) God is perfect.

Jude (Latin) Right in the law.

Jules (Greek) Youthful, downy-bearded.

Julian (22/4) (Latin) Related to Julius.

Julio (22/4) (Greek) Youthful, downy-bearded.

Junius (22/4) (Latin) Born in June.

Kaleb (Hebrew) Spontaneous.

Kane (Gaelic) Tribute.

Kearn (22/4) (Gaelic) Dark.

Kenith (Gaelic) Handsome.

Kenji (22/4) (Japanese) Intelligent second son.

Kennard (Gaelic) Brave chieftain.

Kermit (Gaelic) Free man.

Kieran (Gaelic) Little and dark-skinned.

Kirk (22/4) (Old Norse) From the church.

Kristopher (Greek) Bearer of Christ.

Langdon (Old English) From the long hill.

Langley (Old English) From the long meadow.

Laszlo (22/4) (Hungarian) Famous leader.

Lawton (22/4) (Old English) From the estate on the hill.

Lemar (22/4) (French) Of the sea.

Lionel (Old French) Young lion.

Logan (22/4) (Scottish Gaelic) Little hollow.

Louis (22/4) (Old German) Famous warrior.

Lucius (22/4) (Latin) Light.

Ludwig (Old German) Famous warrior.

Luke (Latin) Light.

Lundy (22/4) (Old French) Born on Monday.

Mace (Old French) Stoneworker.

Makani (22/4) (Hawaiian) Wind.

Malin (22/4) (Old English) Little warrior.

Mandel (22/4) (German) Almond.

Maynard (Old German) Powerful, brave.

Merritt (Old English) Worthy.

Merton (Old English) Town by the sea.

Miguel (Hebrew) Like unto the Lord.

Miles (22/4) (Latin) Soldier, warrior.

Milo (22/4) (Latin) Miller.

Monte (22/4) (Latin) From the pointed mountain.

Morey (Latin) Dark-skinned.

Murdock (Scottish Gaelic) Wealthy sailor.

Myron (Greek) Fragrant ointment.

Nate (Hebrew) Gift of God.

Nathan (22/4) (Hebrew) Gift of God.

Neil (22/4) (Gaelic) The champion.

Nixon (Greek) The people's victory.

Normy (Old French) A northman.

Northrup (Old English) From the north farm.

Obadiah (Hebrew) Servant of God.

Octavio (Latin) The eighth.

Ossie (22/4) (Old English) Divine spearman.

Otho (22/4) (Old German) Rich.

Paget (22/4) (French) Useful assistant.

Patric (Latin) Wellborn, noble.

Pedro (Latin) Rock or stone.

Pembroke (Celtic) From the headland.

Penn (22/4) (Old English) Enclosure.

Percy (Old French) Valley-piercer.

Piers (Latin) Rock or stone.

Putnam (22/4) (Old English) Dweller by the pond.

Quintus (Latin) Fifth Child.

Raff (22/4) (Old English) Swift wolf.

Raoul (22/4) (Old English) Swift wolf.

Redd (22/4) (Old English) Red-colored.

Regis (Latin) Rules.

Remus (22/4) (Latin) Fast-moving.

Reuven (Hebrew) Behold a son.

Reynaldo (Old English) Mighty and powerful.

Reynard (Old French) Fox.

Ripley (Old English) From the shouter's meadow.

Rodger (Old German) Famous spearman.

Rodolfo (Old German) Famous wolf.

Rory (Gaelic) Red king.

Roy (22/4) (Old French) King.

Rudolf (Old German) Famous wolf.

Rudolph (Old German) Famous wolf.

Rufus (22/4) (Latin) Red-haired.

Russel (22/4) (Old French) Red-haired.

Sewell (22/4) (Old English) Sea powerful.

Shelden (Old English) From the town on the ledge.

Siggy (Old German) Victorious protector.

Sinclair (Old French) From Saint Clair.

Skipper (Scandinavian) Shipmaster.

Solomon (Hebrew) Peaceful.

Sonnie (English) Son.

Stanton (22/4) (Old English) From the rocky farm.

Stanway (22/4) (Old English) From the rocky road.

Stein (22/4) (Old German) Rock.

Stern (22/4) (Old English) Austere.

Steven (22/4) (Greek) Crowned.

Taj (Hindi) Crown.

Taji (Japanese) Silver and yellow color.

Teo (Greek) Gift of God.

Terran (English) Earth man.

Terrill (Old German) Martial.

Theobald (Old German) Ruler of the people.

Thomas (22/4) (Hebrew) The devoted brother.

Torin (Gaelic) Chief.

Torrance (Gaelic) From the knolls.
Travers (Old French) At the crossroads.
Tremaine (Celtic) From the house of stone.
Tyrus (22/4) (Old Norse) Thunder.
Valentino (Latin) Strong.
Waite (22/4) (Old English) Guard.
Wakefield (Old English) From the wet field.
Wald (Old German) Ruler.
Walton (22/4) (Old English) From the walled town.
Washington (Old English) From the astute one's estate.
Waylan (22/4) (Old English) From the land by the road.
Welby (22/4) (Old German) From the farm by the spring.

Wheeler (Old English) Wheel-maker.
Wilbur (German-French) Brilliant hostage.
Wilden (Old English) From the wooded hill.
Wolfgang (Old German) Advancing wolf.
Woodman (Old English) From the passage in the woods.
Wright (Old English) Carpenter.
Wynn (22/4) (Old English) From the friendly place.
Xeno (22/4) (Greek) Strange voice.
Yvon (22/4) (Old French) Archer with a yew bow.
Zebulen (Hebrew) Dwelling place.

NUMBERS FOUR AND TWENTY-TWO NAMES—UNISEX

Ali (Arabic) Greatest.
Arnelle (Old German) Strong as an eagle.

Aubry (22/4) (Old French) Blond ruler.
Billie (Old German) Determined guardian.

Blake (Old English) Dark hair and complexion.

Brandee (Dutch) Brandy drink.

Carol (22/4) (Old German) Noble and strong.

Codi (22/4) (Old English) A cushion.

Cord (22/4) (Old French) Ropemaker.

Courtney (Old English) From the court.

Cydney (Old French) From Saint Denis.

Dale (Old English) From the valley.

Dallas (Gaelic) Wise.

Danny (22/4) (Hebrew) God is my judge.

Darcie (Old French) From the fortress.

Darel (22/4) (Old French) Beloved.

Evelin (Hebrew) Life-giving.

Garnett (Old English) Garnet gem.

Gene (22/4) (Greek) Wellborn noble.

Germain (Old German) People of the spear.

Hayley (Old English) Hay meadow.

Hillary (Latin) Cheerful.

Jerry (Latin) Sacred or holy name.

Jesse (Hebrew) Wealthy.

Jessie (22/4) (Hebrew) Wealthy.

Joan (Hebrew) God's gift of grace.

Kiki (22/4) (Old German) Lord of the manor.

Kit (Greek) Bearer of Christ.

Lee (Old English) From the meadow.

Linden (Old English) From the linden tree.

Marlin (Old English) Falcon or hawk.

Renny (Latin) Reborn.

Rikki (Old German) Wealthy, powerful.

Robin (Old English) Bright fame.

Rusty (22/4) (French) Red-haired.

Ryan (22/4) (Gaelic) Little king.

Ryley (Gaelic) Valiant.

Shannon (Gaelic) Small and wise.

Sidney (Old French) From Saint Denis.

Storm (22/4) (English) Storm.

Teddy (22/4) (Greek) Gift of God.

Tobey (22/4) (Hebrew) The Lord is good.

Tracy (22/4) (Gaelic) Battler.

Vale (Old English) From the valley.

Winny (Old German) Peaceful friend.

9

number Five

Qualities of number Five

Keywords: adventurous, clever, adaptable, freedom-loving, communicative, traveler, intellectual, cerebral, flexible, literary, charming, jolly, easygoing, versatile, good-natured, resourceful, sensual, clever, original, resilient, alert, flighty, irritable, uncertain, wasteful, risk-taker, sarcastic, impulsive, hedonistic, shallow, undependable, restless, critical, conceited, nervous, promiscuous.

Symbolism: The number five represents change. Five is masculine in nature and stands for adaptability. Its symbols are Mercury and the pentagram and its element is air. Five is related to the Hermes magician; to humans, who have two arms, two legs, and a head; to the five senses; to the five elements of Chinese philosophy; and to the "Quintessence" (Fifth Essence), which is an alchemical term meaning the four elements plus physical consciousness.

Personality: The number five person is a freedom-loving adventurer who adapts easily to any new situation. Five's greatest talent is the ability to cope with the unexpected. The Five loves to try anything that is new, exciting, and unusual. Taking chances is second nature to Fives, and they are always ready to gamble on the next big deal. Fives live on nervous energy and are constantly on the move. They have an inborn desire for travel and adventure. The word ''fear'' is not included in their vocabulary, thus they seem to be indifferent to danger. Fives have inquiring minds and need to know why things are the way they are. Their lives are filled with events, often of short duration. Fives are good mixers and make friends easily. Because of a great elasticity of character a Five will never remain down for long. They tend to be lucky and are able to bounce back from almost any difficulty.

Fives are sensual and sexy and can also be loving and affectionate, but they are likely to jump in and out of romantic relationships. Although they will rarely admit to being fickle, there is a fear of intimacy and a tendency to move on. They can be critical and may hurt others with sarcastic remarks. If they become excessively restless, they can become mentally unstable.

Careers: astrologer, aviator, broker, civic leader, detective, editor, explorer, gambler, instructor, importer/exporter, lecturer, lobbyist, musician, promoter, racing driver, reporter, salesperson, speculator, stockbroker, travel agent, wholesaler.

NUMBER FIVE NAMES—FEMALE

Abigail (Hebrew) My father is joy.

Addie (Old German) Of noble rank.

Adela (Old German) Of noble rank.

Adelaide (Old German) Of noble rank.

Adeline (Old German) Of noble rank.

Alair (Latin) Cheerful, glad.

Alberta (Old English) Noble and brilliant.

Aleda (Greek) Beautifully dressed.

Alfonsine (Old German) Noble and eager.

Alia (Israeli) Immigrant to a new home.

Aline (Old German) Of noble rank.

Alisabeth (Hebrew) God's promise.

Alisha (Greek) Truthful.

Allaryce (Old German) Rules all.

Alphonsia (Old German) Noble and eager.

Alvina (Old German) Beloved by all.

Alyssa (Greek) Truthful.

Amapola (Arabic) Poppy.

Amarantha (Greek) Beyond death.

Amaya (Japanese) Night rain.

Amelia (Old German) Industrious.

Amelinda (Spanish) Beloved and pretty.

Amity (Latin) Friendship.

Anya (Hebrew) Full of grace.

Apollonia (Latin) Belonging to Apollo.

Aracelly (Latin) Altar of heaven.

Astra (Latin) Star.

Auberta (Old English) Noble and brilliant.

Azusa (Arabic) Lily.

Becca (Hebrew) The captivator.

Bella (French) Beautiful.

Besse (Hebrew) God's promise.

Bessie (Hebrew) God's promise.

Bethea (Hebrew) Maidservant of Jehovah.

Blodwyn (Welsh) White flower.

Blossom (Old English) Flowerlike.

Bonnie (Scottish-English) Beautiful, pretty.

Brianna (Celtic) Strength with virtue and honor.
Brigitta (Celtic) Strong and mighty.
Briona (Celtic) Strength with virtue and honor.
Brita (Latin) From England.
Britteny (Latin) From England.
Brittney (Latin) From England.
Caitlin (Greek) Pure maiden.
Candra (Latin) Luminescent.
Cara (Latin) Dear.
Carly (Old German) Little woman born to command.
Caroline (Old German) Little woman born to command.
Ceres (Latin) Of the spring.
Cerise (French) Cherry, cherry red.
Chantal (Latin) Song.
Chantilly (French) Lace from Chantilly.
Chelsey (Old English) A port of ships.
Cherry (English) Cherry fruit.
Ciandra (Hebrew) God's gracious gift.
Ciara (Gaelic) Little and dark-skinned.
Cleta (Greek) Summoned.
Consuelo (Spanish) Consolation.
Corliss (Old English) Cheerful, good-hearted.

Corneila (Latin) Yellow, horn-colored.
Crissi (Greek) Anointed.
Danica (Slavic) The morning star.
Darlene (Old French) Little darling.
Dee (Welsh) Black, dark.
Deva (Sanskrit) Divine.
Divina (Latin) Divine.
Dolly (Greek) Gift of God.
Dominica (Latin) Belonging to the Lord.
Dorothea (Greek) Gift of God.
Duana (Gaelic) Dark.
Dulcine (Latin) Sweet and charming.
Easter (Old English) Easter time.
Edda (Old German) Strives.
Edie (Old English) Rich gift.
Eileen (Greek) Light, a torch.
Elise (Hebrew) God's promise.
Elna (Greek) Light, a torch.
Elsie (Hebrew) God's promise.
Emma (Old German) One who leads the universe.
Enid (Old Welsh) Purity.
Enrica (Old German) Lord of the manor.
Estefany (Greek) Crowned.
Ethel (Old English) Noble.
Eve (Hebrew) Life-giving.
Evelina (Hebrew) Life-giving.

Fatima (Arabic) Wean or abstain.

Faustine (Latin) Fortunate.

Fay (Old German) Belief in God.

Fidelma (Latin-Hebrew) Faithful, bitter.

Filma (Old English) Misty veil.

Finella (Gaelic) White-shouldered.

Fionna (Gaelic) Fair.

Fortuna (Latin) Fortunate.

Gabriele (Hebrew) Devoted to God.

Gaea (Greek) The earth.

Gardenia (Scottish) Gardenia flower.

Gayle (Old French) Bright and lively.

Genna (French) Pure white wave.

Georgiana (Greek) Landholder, farmer.

Germana (Old German) People of the spear.

Gigi (Old German) Brilliant hostage.

Ginamarie (Latin-Hebrew) Queen, bitter.

Gladys (Celtic) Princess.

Gloriana (Latin) Glory.

Glory (Latin) Glory.

Glynis (Old Welsh) Dweller in a valley.

Haidee (Greek) Well-behaved.

Hanako (Japanese) Flower child.

Hortense (Latin) Gardener.

Hylda (Old German) Battle maid.

Ida (Old German) Industrious.

Idelisa (Celtic) Bountiful.

Ilsa (Hebrew) God's promise.

Imogene (Latin) Image.

Ingeborg (Norse) Hero's daughter.

Irma (Latin) High-ranking person.

Ishmaela (Hebrew) God will hear.

Iva (Old French) Yew tree.

Jacoba (Hebrew) The supplanter.

Janet (Hebrew) God's gift of grace.

Janthina (Greek) Violet-colored flower.

Jarita (Arabic) Earthen water jug.

Jenny (Hebrew) God's gift of grace.

Jeovana (Hebrew) God's gracious gift.

Jessamine (Arabic) Fragrant flower.

Jilly (Greek) Young in heart and mind.

Joanne (Hebrew) God's gift of grace.

Joelle (Hebrew) Jehovah is God.

Josilyn (Latin) Fair and just.

Jovita (Latin) Father of the sky.

Joy (Latin) Rejoicing.

Juliana (Greek) Young in heart and mind.

Juliet (Greek) Young in heart and mind.

June (Latin) Young.

Kady (Greek) Pastoral simplicity and happiness.

Kallie (Scandinavian) Flowing water.

Kanani (Hawaiian) The beautiful one.

Katha (Greek) Pure maiden.

Katleen (Greek) Pure maiden.

Kayla (Greek) Pure maiden.

Kaylee (Greek) Pure maiden.

Kelila (Hebrew) Crown, laurel.

Kimiko (Japanese) Heavenly child.

Kiona (American Indian) Brown hills.

Kitra (Hebrew) Crowned.

Kyoko (Japanese) Mirror.

Laine (Old English) From the narrow road.

Lara (Latin) Shining, famous.

Latrice (Latin) Happiness.

Laverne (Latin) Springlike.

Lavina (Latin) Purified.

Lavinia (Latin) Purified.

Laylah (Arabic) Dark as night.

Lena (Greek) Light.

Leorah (Old French) Light.

Leticia (Latin) Gladness.

Libby (Hebrew) Consecrated of God.

Lilibeth (Latin-Hebrew) Lily flower, house of God.

Lille (Latin) Lily flower.

Lillie (Latin) Lily flower.

Lisa (Hebrew) God's promise.

Lorrie (Latin) Crowned with laurels.

Louisa (Old German) Famous battle maid.

Luz (Spanish) Light.

Madeleine (Hebrew) Woman of Magdala.

Maemi (Japanese) Smile of truth.

Mai (Latin) Great.

Mame (Hebrew) Bitterness.

Mamie (Hebrew) Bitterness.

Mandi (Latin) Worthy of being loved.

Margaretta (Latin) A pearl.

Marguerita (Latin) A pearl.

Mari (Hebrew) Bitterness.

Mariah (Hebrew) Bitterness.

Maribelle (Hebrew-French) Bitter, beautiful.

Mariela (Hebrew) Bitterness.

Marisel (Latin) Of the sea.

Marlene (Hebrew) Bitterness.

Maryann (Hebrew) Bitter, graceful.

Mathilda (Old German) Powerful in battle.
Maureen (Hebrew) Bitterness.
Maurene (Hebrew) Bitterness.
Meagan (Gaelic) Soft, gentle.
Media (Greek) Ruling.
Melanie (Greek) Dark-clothed.
Merrie (English) Mirthful, joyous.
Mia (Latin) Mine.
Michiko (Japanese) Beauty and wisdom.
Midori (Japanese) Green.
Minette (Old German) Love, tender affection.
Mira (Latin) Wonderful.
Misty (Old English) Covered with mist.
Mitzi (Hebrew) Bitterness.
Molly (Hebrew) Bitterness.
Myrta (Greek) Myrtle plant.
Nanci (Hebrew) Full of grace.
Nanny (Hebrew) Full of grace.
Nariko (Japanese) Thunderpeal.
Nathania (Hebrew) Gift of God.
Neila (Gaelic) The champion.
Nelia (Latin) Yellow, horn-colored.
Nelly (Old French) Light.
Nesta (Greek) Pure.
Nissy (Norse) Friendly elf or brownie.
Noella (French) Born at Christmas.

Odele (Hebrew) I will praise God.
Olida (Latin) Symbol of peace.
Olivia (Latin) Symbol of peace.
Omega (Greek) Ultimate.
Opalina (Sanskrit) A precious stone.
Pammy (Greek) All honey.
Paolina (Latin) Little.
Patrica (Latin) Wellborn, noble.
Patricia (Latin) Wellborn, noble.
Petunia (American Indian) Petunia flower.
Preciosa (Old French) Of great value.
Primrose (Latin) First rose.
Prissie (Latin) From ancient times.
Prudence (Latin) Foresight.
Rachell (Hebrew) Innocent as a lamb.
Ramah (Israeli) High.
Rayna (Sanskrit) Queen.
Rebekah (Hebrew) The captivator.
Renata (Latin) Reborn.
Rhea (Greek) Flows from the earth.
Rima (Spanish) Rhyme, poetry.
Riva (French) Riverside.
Robina (Old English) Bright fame.
Rosalba (Latin) White rose.
Rosalyn (Spanish) Beautiful rose.

Rosanne (Greek-Hebrew) A rose, full of grace.
Rosy (Greek) A rose.
Roya (Old French) Royal.
Sabra (Hebrew) Thorny cactus.
Samantha (Aramaic) Listener.
Sarita (Hebrew) Of royal status.
Selima (Hebrew) Peaceful.
Selma (Old Norse) Divinely protected.
Shanice (African) Marvelous.
Sheela (Celtic) Musical.
Sherri (Old French) Beloved.
Siobhan (Hebrew) Admired.
Sofia (Greek) Wisdom.
Sonja (Greek) Wisdom.
Sophia (Greek) Wisdom.
Sorelle (Old French) Reddish-brown.
Tabby (Greek) Gazelle.
Taja (Hindi) Crown.
Teresa (Greek) The harvester.
Tessie (Greek) The harvester.
Thelma (Greek) A nursling.
Theodora (Greek) Gift of God.
Thomasine (Hebrew) The devoted sister.
Tiphani (Greek) Appearance of God.

Tita (Latin) Title of honor.
Tonia (Latin) Praiseworthy, without a peer.
Trilby (Italian) To sing with trills.
Trude (Old German) Beloved.
Tuesday (Old English) Tuesday.
Udelle (Old English) Prosperous.
Valera (Latin) Strong.
Valeria (Latin) Strong.
Veda (Sanskrit) Wise.
Velvet (Old English) Velvet.
Verda (Latin) Young, fresh.
Veronika (Latin) Springlike, youthful.
Valarie (Latin) Strong.
Viola (Latin) Violet flower.
Xantha (Greek) Golden-yellow.
Xanthippe (Greek) Yellow horse.
Yasmine (Arabic) Jasmine flower.
Yoninah (Hebrew) Dove.
Yvonne (Old French) Archer with a yew bow.
Zaida (Arabic) Good fortune.
Zillah (Hebrew) Shadow.
Zohra (Arabic) Blossom.

NUMBER FIVE NAMES—MALE

Alberic (Old German) Elf ruler.
Aldo (Latin) Elder.
Algernon (Old French)
Mustached or bearded.
Alvan (Old German) Beloved by
all.
Alwin (Old German) Beloved by
all.
Amir (Arabic) Prince.
Anatole (Greek) From the east.
Arlen (Gaelic) A pledge.
Arnaud (Old German) Strong as
an eagle.
Arthur (Celtic-Welsh) Noble,
bear-hero.
Ayers (Old English) Heir to a
fortune.
Baron (Old English) Nobleman,
baron.
Bart (Hebrew) Son of a farmer.
Benjamin (Hebrew) Son of my
right hand.
Bernardo (Old German) Brave
as a bear.
Bertie (Old English) Bright.
Bertram (Old English) Brilliant
raven.
Bing (Old German) Kettle-
shaped hollow.
Bjorn (Old German) Bear.

Bowen (Welsh) Son of Owen.
Boyce (Old French) From the
woodland.
Bradford (Old English) From
the broad river crossing.
Brady (Old English) From the
broad island.
Bramwell (Old English) At
Bram's well.
Brandon (Old English) From the
beacon hill.
Brandt (Old English) Firebrand.
Brennan (Gaelic) Little raven.
Brent (Old English) Steep hill.
Burdette (Old French) Little
shield.
Cable (Old English) Ropemaker.
Caine (Hebrew) Possession or
possessed.
Caleb (Hebrew) Spontaneous.
Carlos (Old German) Strong
and masculine.
Carne (Gaelic) Victorious.
Cashman (Latin) Vain.
Cecil (Latin) Dim-sighted or
blind.
Chasen (Old French) Hunter.
Chic (Old German) Strong and
masculine.
Clarke (Old French) Scholar.

Claud (Latin) Lame.

Clay (Old English) From the place of clay.

Clayborne (Old English) Born of clay.

Cletis (Greek) Famous.

Colt (English) Young horse.

Conlan (Gaelic) Hero.

Connal (Celtic) Mighty, valorous.

Cyrus (Old Persian) The sun.

Dave (Hebrew) Beloved.

Davie (Hebrew) Beloved.

Denver (Old English) Green valley.

Derick (Old German) Leader of the people.

Derrick (Old German) Leader of the people.

Desiderus (Latin) Yearning, sorrow.

Domingo (Latin) The Lord's day.

Donahue (Gaelic) Brown-haired fighter.

Donald (Gaelic) World ruler.

Dov (Hebrew) Beloved.

Drury (Old French) Beloved friend.

Duke (Old French) Leader, duke.

Earle (Old English) Nobleman.

Eder (Hebrew) Flock.

Edrick (Old English) Power and good fortune.

Eduardo (Old English) Prosperous guardian.

Egon (Gaelic) Ardent, fiery.

Eldon (Old English) Wise protector.

Eleazar (Hebrew) God has helped.

Ellery (Old English) From the elder tree island.

Ellison (Old English) Son of Ellis.

Errol (Old English) Nobleman.

Ervin (Old English) Sea friend.

Everett (Old English) Strong as a boar.

Evin (Gaelic) Wellborn young warrior.

Ezra (Hebrew) Helper.

Fabrizio (Latin) Craftsman.

Fergie (Gaelic) The best choice.

Fernando (Old German) World-daring.

Fiske (Old English) Fish.

Fletcher (Old English) Arrow-featherer.

Fonsie (Old German) Noble and eager.

Frank (Latin) Free man.

Frederic (Old German) Peaceful ruler.

Gareth (Welsh) Gentle.

Geno (Hebrew) God's gracious gift.

Geomar (Old German) Famous in battle.

Gerardo (Old English) Spear-strong.

Gervaise (Old German) Keen with a spear.

Gervase (Old German) Honorable.

Gilroy (Gaelic) Devoted to the king.

Gonsalve (Old German) Complete warrior.

Gordan (Old English) Round hill.

Gregory (Latin) Watchman.

Grenville (Old French) From the large town.

Gustave (Swedish) Staff of the Goths.

Gyles (Greek) Shield bearer.

Hadrien (Latin) From Adria.

Hadwin (Old English) Battle companion.

Hamar (Old Norse) Symbol of ingenuity.

Hamlet (Old French-German) Little home.

Hammond (Old German) Small home lover.

Hansel (Hebrew) God's gracious gift.

Harcourt (French) Fortified dwelling.

Harlow (Old English) From the rough hill.

Havelock (Old Norse) Sea battle.

Heinrick (Old German) Lord of the manor.

Herman (Latin) High-ranking person.

Heywood (Old English) From the hedged forest.

Homer (Greek) Promise.

Horace (Latin) Keeper of the hours.

Horatio (Latin) Keeper of the hours.

Howland (Old English) From the hills.

Hughes (Old English) Intelligence.

Hunter (Old English) A hunter.

Hussein (Arabic) Little and handsome.

Huxley (Old English) From Hugh's meadow.

Iago (Hebrew) The supplanter.

Isaak (Hebrew) He laughs.

Ismael (Hebrew) God will hear.

Ivar (Old Norse) Battle archer.

Jabari (Swahili) Valiant.

Jason (Greek) Healer.

Jedidiah (Hebrew) Beloved of the Lord.

Jerald (Old German) Spear-mighty.

Jericho (Arabic) City of the moon.

Jim (Hebrew) The supplanter.

Joachim (Hebrew) The Lord will judge.

Jonas (Hebrew) Dove.

Josephus (Hebrew) He shall add.

Kaemon (Japanese) Joyful.

Kameron (Scottish Gaelic) Crooked nose.

Keefe (Gaelic) Cherished, handsome.

Keefer (Gaelic) Cherished, handsome.

Keenan (Gaelic) Little and ancient.

Kelton (Old English) Town where ships are built.

Kenneth (Gaelic) Handsome.

Kent (Old Welsh) White, bright.

Kevyn (Gaelic) Gentle, kind, and lovable.

Killian (Gaelic) Little and warlike.

King (Old English) King.

Klinton (Old English) From the headland farm.

Kordell (Old French) Ropemaker.

Lars (Latin) Crowned with laurels.

Leander (Greek) Like a lion.

Leif (Old Norse) Beloved.

Lemuel (Hebrew) Consecrated to God.

Lennie (Old French) Lion-brave.

Leo (Latin) Lion.

Lewis (Old German) Famous warrior.

Lindon (Old English) From the linden tree.

Lloyd (Old Welsh) Grey-haired.

Lon (Old German) Noble and eager.

Ludovic (Old German) Famous warrior.

Macnair (Scottish Gaelic) Son of the heir.

Marcellus (Latin) Follower of Mars.

Marco (Latin) Follower of Mars.

Marsh (Old French) Steward, horse-keeper.

Marvin (Old English) Lover of the sea.

Medwin (Old German) Strong friend.

Merrick (Old English) Ruler of the sea.

Miner (Latin) Junior, younger.

Mordecai (Hebrew) Warrior.

Morland (Old English) Marsh, wetland.

Morton (Old English) From the town on the moor.

Munroe (Gaelic) From the mouth of the Roe River.

Murry (Scottish Gaelic) Sailor.

Neal (Gaelic) The champion.

Ned (Old English) Rich and happy protector.

Nels (Gaelic) The champion.

Neuman (Old English) New arrival.

Niels (Gaelic) The champion.

Niles (Gaelic) The champion.

Noach (Hebrew) Rest, comfort.

Ogdan (Old English) From the oak valley.

Olav (Old Norse) Ancestral talisman.

Ole (Old Norse) Ancestral talisman.

Olin (Old Norse) Ancestral talisman.

Orel (Russian) Eagle.

Orval (Old English) Mighty spear.

Osbourn (Old English) Warrior of God.

Osmund (Old English) Divine protector.

Oswell (Old English) Divinely powerful.

Otes (Greek) Keen of hearing.

Paddy (Latin) Wellborn, noble.

Paolo (Latin) Little.

Patton (Old English) From the warrior's estate.

Paul (Latin) Little.

Pembrook (Celtic) From the headland.

Percival (Old French) Valley-piercer.

Peyton (Old English) From the warrior's estate.

Procter (Latin) Administrator.

Quillan (Gaelic) Cub.

Quintin (Latin) Fifth child.

Rad (Old English) From the red stream.

Raddie (Old English) From the red stream.

Ram (Old English) From the Ram's island.

Rami (Arabic) Loving.

Ramsay (Old English) From the ram's island.

Rance (African) Borrowed all.

Randal (Old English) Shield-wolf.

Rashid (Arabic) Counselor.

Ravi (Hindi) Sun.

Reed (Old English) Red-haired.

Rei (Japanese) Law, rule.

Reinhard (Old French) Fox.

Ricardo (Old German) Wealthy, powerful.

Rick (Old German) Wealthy, powerful.

Roald (Scandinavian) Renowned, powerful.

Roarke (Gaelic) Famous ruler.

Rodd (Old English) One who rides with a knight.

Roosevelt (Old Dutch) From the rose field.

Roswell (Old English) From a field of roses.

Rudy (Old German) Famous wolf.

Rurik (Old German) Famous ruler.

Russ (Old French) Red-haired.

Ryker (Old German) Wealthy, powerful.

Sage (Old French) Wise one.

Salvatore (Italian) Savior.

Sanford (Old English) From the sandy hill.

Sanjiro (Japanese) Admirable.

Santiago (Spanish) Saint James.

Sayer (Welsh) Carpenter.

Sayre (Welsh) Carpenter.

Scott (Old English) Scotsman.

Septimus (Latin) Seventh-born son.

Serle (Old German) Bearer of arms.

Shalom (Hebrew) Peace.

Sheldon (Old English) From the town on the ledge.

Sid (Old French) From Saint Denis.

Sigfried (Old German) Victorious protector.

Silvio (Latin) From the forest.

Skell (Gaelic) Storyteller.

Skipton (Scandinavian) Shipmaster.

Slade (Old English) Child of the valley.

Stanleigh (Old English) From the rocky meadow.

Steinar (Old German) Rock warrior.

Sterling (Old English) Valuable.

Sutherland (Scandinavian) From the southern land.

Tab (Old English) Drummer.

Teague (Gaelic) Bard.

Thayer (Old French) From the nation's army.

Thorndike (Old English) From the thorny embankment.

Toddy (Old English) A fox.

Tomas (Hebrew) The devoted brother.

Tomeo (Japanese) Cautious man.

Tommy (Hebrew) The devoted brother.

Toshiro (Japanese) Talented.

Trent (Latin) Torrent.

Trey (Old English) Three, the third.

Upton (Old English) From the upper town.

Vern (Latin) Springlike, youthful.

Virgil (Latin) Staff-bearer.

Waldemar (Old German) Ruler.

Walden (Old English) From the woods.

Wayne (Old English) Wagoner.

Webb (Old English) Weaver.

Whitaker (Old English) From the white field.

Wildon (Old English) From the wooded hill.

Wilfred (Old German) Resolute and peaceful.

Woodrow (Old English) From the passage in the woods.
Xenos (Greek) Stranger.
Xerxes (Persian) Ruler.
Yancy (American Indian) Englishman.
Zaccheus (Hebrew) Jehovah has remembered.
Zacharias (Hebrew) Jehovah has remembered.

Zachery (Hebrew) Jehovah has remembered.
Zander (Greek) Helper of humankind.
Zebulon (Hebrew) Dwelling place.
Zelig (Old German) Blessed.

NUMBER FIVE NAMES—UNISEX

Ashton (Old English) From the ash tree town.
Blayne (Gaelic) Thin, lean.
Courtenay (Old English) From the court.
Cruz (Spanish) Cross.
Cyd (Old French) From Saint Denis.
Darby (Gaelic) Free man.
Drew (Latin) Strong.
Haleigh (Old English) Hay meadow.
Haven (Old English) Place of safety.
Jacky (Hebrew) The supplanter.
Jaye (Old French) Blue jay.

Johnny (Hebrew) God's gracious gift.
Kellen (Gaelic) Warrior.
Kelsey (Old Norse) From the ship island.
Kendall (Old English) From the bright valley.
Kerry (Gaelic) Dark one.
Kimberly (Old English) From the king's wood.
Lane (Old English) From the narrow road.
Lanie (Old English) From the narrow road.
Leigh (Old English) From the meadow.

Loni (Old English) Solitary.

Lorin (Latin) Crowned with laurels.

Lyndsey (Old English) From the linden tree island.

Marlo (Old English) From the hill by the lake.

Marty (Latin) Follower of Mars.

Mattie (Hebrew) Gift of God.

McKenzie (Gaelic) Son of the wise leader.

Mead (Old English) From the meadow.

Meridith (Old Welsh) Guardian from the sea.

Michel (Hebrew) Like unto the Lord.

Morgan (Scottish Gaelic) From the edge of the sea.

Reagen (Gaelic) Little king.

Rockey (Old English) From the rock.

Ronny (Old English) Mighty and powerful.

Rosario (Spanish) Rosary.

Sacha (Greek) Helper of humankind.

Sal (Italian) Savior.

Sandye (Old English) From the sandy hill.

Shelley (Old English) From the meadow on the ledge.

Skylar (Dutch) Scholar.

Stacy (Latin) Stable, prosperous.

Terry (Greek) The harvester.

Tonnie (Latin) Praiseworthy, without a peer.

Vivian (Latin) Alive.

Whitney (Old English) From the white island.

10

Number Six/Thirty-Three

Qualities of Number Six

Keywords: loving, artistic, caring, compassionate, magnetic, affectionate, concerned, responsible, refined, sensitive, dependable, balanced, thoughtful, stable, sociable, outgoing, kind, consistent, parental, domestic, harmonious, loyal, honest, logical, fortunate, cheerful, peaceful, vain, lazy, gossipy, complacent, interfering, meddlesome, jealous, self-righteous, obstinate, dogmatic, stubborn, egotistical, cynical.

Symbolism: The number six represents harmony. Six is feminine in nature and stands for balance. Its symbols are Venus and the double triangle (six-pointed star), and its element is earth. Six is sometimes referred to as a perfect number, because it equals the sum of its divisors (1+2+3=6) and is divisible by both the odd number 3 and the even number 2, thus harmoniously combining the elements of each. It is related to the human soul, love, marriage, domestic happiness, the goddess Diana, and the heart.

Personality: The number six personality is creative, artistic, imaginative and often drawn to the arts as a means of self-expression. Sixes are romantically inclined and are very attractive to the opposite sex. Because they are in love with love, their outlook on relationships is decidedly idealistic. Sixes have magnetic personalities and tend to draw people to them. They love their homes and derive much pleasure from being surrounded by beautiful objects. Sixes love to entertain and will spend lavishly for parties and other social affairs. Considered very fortunate, they tend to attract money and property, and their partnerships and business investments are generally lucrative. Sixes, however, place more importance on people and pleasure than on material possessions. They refuse to live as hermits, and love, sex, and marriage are necessary for their completion. Sixes make excellent parents, and many have large families. They strive for harmony, and they like their lives to run smoothly, with a minimum of stress and discord.

Their concern for fellow humans and their love of justice can be taken too far, and Sixes are often accused of meddling in the affairs of others. The desire for beauty and pleasant surroundings can become so exaggerated that it becomes fussy and trivial, and there can be a tendency toward vanity, laziness, and gossip. Sixes can become petty or childish when their feelings are hurt.

Careers: artist, athlete, barber, beautician, cashier, cosmetician, dentist, doctor, educator, florist, homemaker, interior designer, jeweler, judge, lawyer, musician, nurse, personal trainer, professor, realtor, singer, social worker, surgeon, teacher, waiter.

Qualities of Number Thirty-Three (A master number and the higher form of six)

Keywords: emotional, ecstatic, loving, caring, charming, devotional, self-sacrificing, compassionate, concerned, magnetic, sensitive, balanced, sociable, disciplined, enthusiastic, dramatic, inspirational, powerful, hostile, confused, chaotic, erratic, disruptive, unfeeling, controlling.

Symbolism: The number thirty-three is a master number, and as the higher form of six it represents emotional mastery. Thirty-three is made up of the forces of eleven (spiritual) and twenty-two (physical) and is considered by some to stand for the Savior. The positive thirty-three person is an idealistic leader with the power to command or serve. Those who are able to live up to this intense vibration know how to control their emotions without repressing them. Thus they are both masters of their emotions and emotional masters.

Personality: The Thirty-three has come into this life to demonstrate mastery over human emotions and to pass this knowledge on to others who can then live a freer life. They understand that our emotions are reactions to experiences and that when the causes of these experiences are understood we can erase negative programming and reverse the directions of our lives. With diligence and study Thirty-threes learn how to master their own emotions and understand the emotional makeup of others. Their task is to work through feelings, their own and others, in order to bring about agreement. The power to command people by triggering their emotions is very strong with this personality so they must be careful not to abuse it. Through the use of charm and the ability to understand what other people are feeling, Thirty-threes can learn how to sway an entire group one way or another. As orators they easily exert their emotional control over individuals and groups.

Thirty-threes can become addicted to emotions. They may experience the entire emotional spectrum, going from ecstatic raptures to lows of self-denial and self-punishment. The power of Thirty-threes to influence other people's emotions sometimes works to their own detriment. They may spur riots or otherwise disrupt the orderly functions of society. By altogether repressing their feelings they can develop

NUMBERS SIX AND THIRTY-THREE NAMES—FEMALE

Abrianna (33/6) (Hebrew) Mother of the multitude.

Ada (Hebrew) Ornament.

Adabelle (Hebrew-French) Beautiful ornament.

Adelisa (Greek) Truthful.

Adira (Arabic) Strong.

Aida (Old English) Prosperous, happy.

Aidee (Greek) Well-behaved.

Aileene (33/6) (Greek) Light, a torch.

Aimee (Latin) The beloved.

Akilah (Arabic) Bright, smart.

Alandra (Greek) Helper of humankind.

Alena (Greek) Light, a torch.

Alisa (Greek) Truthful.

Aloma (Spanish) Dove.

Ama (Ghanese) Saturday's child.

Amabelle (Latin) Lovable.

Ambrosine (Greek) Immortal.

Amira (Arabic) Princess.

Anarosa (Hebrew-Greek) Graceful rose.

Anastacia (Greek) Of the resurrection.

Antoinette (Latin) Praiseworthy, without a peer.

Aphrodite (Greek) Goddess of love and beauty.

Ardis (Hebrew) Flowering field.

Ardith (33/6) (Hebrew) Flowering field.

Arleigh (Old German) Meadow of the hare.

Arliss (Old German) Womanly strength.

Arthurine (Celtic-Welsh) Noble, bear-heroine.

Ava (Latin) Birdlike.

Avis (Old English) Refuge in battle.

Azusena (Arabic) Lily.

Barbra (Latin) Beautiful stranger.

Benita (Latin) Blessed.

Betina (Hebrew) God's promise.

Blanca (Spanish) White, fair.

Bryna (Celtic) Strength with virtue and honor.

Cady (Greek) Pastoral simplicity and happiness.

Callie (Gaelic) Thin, slender.

Camilla (Latin) Young ceremonial attendant.

Cecilia (33/6) (Latin) Dim-sighted or blind.

Celeste (Latin) Heavenly.

Chastity (33/6) (Latin) Purity.

Chica (Spanish) Small girl.

Christa (33/6) (Greek) Anointed.

Christine (Greek) Anointed.

Clarice (33/6) (Latin-Greek) Most brilliant.

Claudia (Latin) Lame.

Claudine (33/6) (Latin) Lame.

Clemence (33/6) (Latin) Merciful.

Clementina (Latin) Merciful.

Conception (Latin) Conception.

Corabella (33/6) (Greek-French) Beautiful maiden.

Corina (33/6) (Greek) The maiden.

Corinne (Greek) The maiden.

Cosette (French) Little pet.

Cybil (Greek) A prophetess.

Dacy (Gaelic) Southerner.

Dara (Hebrew) Compassion.

Daria (Greek) Queenly.

Davina (Hebrew) Beloved.

Dawn (Old English) The break of day.

Delfina (33/6) (Greek) Calmness.

Delilah (33/6) (Hebrew) Delicate, languishing, amorous.

Delta (Greek) Fourth-born.

Dena (Latin) Divine moon goddess.

Desire (33/6) (French) Longed for.

Desma (Greek) A pledge.

Destiny (33/6) (Old French) Destiny.

Devon (Old English) Defender of Devonshire.

Diahann (33/6) (Latin) Divine moon goddess.

Diane (Latin) Divine moon goddess.

Dilys (Welsh) Perfection.

Dinorah (Hebrew) Judged.

Dixie (33/6) (American) Girl of the south.

Dorcas (Greek) Grace of the gazelle.

Dorothy (Greek) Gift of God.

Dorrie (Greek) Gift.

Drusilla (33/6) (Latin) Strong.

Edina (Old English) Rich friend.

Edna (Hebrew) Rejuvenation.

Elana (Hebrew) Tree.

Elfreda (33/6) (Old English) Elfin magic.

Eliana (Greek) Light, a torch.

Elke (Old German) Of noble rank.

Ernestina (Old English) Earnest.

Esme (Old French) Esteemed.

Esmeralda (33/6) (Spanish) Emerald.

Esperanza (Spanish) Hope.

Estelle (Persian) A star.

Euphemia (Greek) Auspicious speech.

Fallon (Gaelic) Grandchild of the ruler.

Fanella (Gaelic) White-shouldered.

Fanny (Latin) Free.

Felisha (33/6) (Latin) Joyous, fortunate.

Flavia (Latin) Yellow-haired.

Flor (Latin) Flower.

Florentina (Latin) Flowering or blooming.

Franny (33/6) (Latin) Free.

Frederica (Old German) Peaceful ruler.

Gay (Old French) Bright and lively.

Gilberte (Old German) Brilliant hostage.

Gilda (Old English) Covered with gold.

Ginger (Latin) Ginger flower or ginger spice.

Ginny (33/6) (Latin) Virginal.

Giordana (Hebrew) Down-flowing.

Giselle (33/6) (Old German) Pledge or hostage.

Greta (Latin) A pearl.

Gwyn (Old Welsh) White-browed.

Hadara (Hebrew) Adorned with beauty.

Hana (Hebrew) Full of grace.

Helga (Norse) Holy.

Heloisa (33/6) (Old German) Famous woman warrior.

Hermione (Greek) Of the earth.

Hertha (33/6) (Old English) Of the earth.

Ileana (Greek) Light, a torch.

Ilona (Greek) Light, a torch.

Imena (African) A dream.

Ina (Greek) Pure.

Iphigenia (Greek) Sacrifice.

Irene (33/6) (Greek) Goddess of peace.

Isolda (Welsh) Fair lady.

Jacinda (Greek) Hyacinth flower.

Janice (Hebrew) God's gift of grace.

Jemima (Hebrew) A dove.

Jemma (Latin) Precious stone.

Jeraldine (Old German) Spear-mighty.

Jeri (Latin) Sacred or holy name.

Jolie (French) Pretty.

Judy (Hebrew) Admired.

Juno (Latin) Wife of Jupiter.

Kachine (33/6) (American Indian) Sacred dancer.

Kali (Sanskrit) Black goddess, energy.

Kallan (Scandinavian) Flowing water.

Karissa (Greek) Very dear.

Karmel (Hebrew) God's vineyard.

Karyn (Greek) Pure maiden.

Kassia (Greek) Sweet-scented spice.

Katharine (Greek) Pure maiden.

Katrine (33/6) (Greek) Pure maiden.

Keiko (Japanese) Adored.

Keziah (33/6) (Greek) Sweet-scented spice.

Kimi (Japanese) Happiness.

Kirsten (33/6) (Norse) Anointed.

Kristen (33/6) (Greek) Anointed.

Kristine (Greek) Anointed.

Kyna (Gaelic) Great wisdom.

Laraine (33/6) (Latin) Crowned with laurels.

Lark (English) Skylark.

Laurel (Latin) Laurel tree.

Lavelle (Latin) Cleansing.

Layla (Arabic) Dark as night.

Lemuela (Hebrew) Dedicated to God.

Leonie (33/6) (Latin) Lion.

Leora (Old French) Light.

Lexine (33/6) (Greek) Helper of humankind.

Lianna (French) A climbing vine.

Liezel (33/6) (Hebrew) God's promise.

Lili (Latin) Lily flower.

Lilibet (33/6) (Hebrew) God's promise.

Lillian (33/6) (Latin) Lily flower.

Lisandra (33/6) (Greek) The liberator.

Lolita (Spanish) Sorrows.

Lomasi (American Indian) Pretty flower.

Lona (Old English) Solitary.

Lorna (Latin) Crowned with laurels.

Lotus (Greek) Lotus flower.

Louella (Old English) Elf.

Louellen (33/6) (Old English) Elf.

Lurleen (33/6) (German) Siren.

Lurlene (33/6) (German) Siren.

Lydia (Greek) A woman from Lydia.

Mabel (Latin) Lovable.

Machiko (33/6) (Japanese) Beautiful child.

Maggie (33/6) (Latin) A pearl.

Maia (Latin) Great.

Majesta (Latin) Royal bearing.

Malana (Hawaiian) Buoyant, light.

Mara (Latin) Of the sea.

Marcy (Latin) Follower of Mars.

Margery (Latin) A pearl.

Maria (Hebrew) Bitterness.

Maribel (33/6) (Hebrew-French) Bitter, beautiful.

Marietta (33/6) (Hebrew)
Bitterness.

Maris (Latin) Of the sea.

Marisela (33/6) (Latin) Of the
sea.

Marisol (33/6) (Hebrew-Latin)
Bitterness, alone.

Marnie (33/6) (Latin) From the
sea.

Marsha (Latin) Follower of
Mars.

Marvina (33/6) (Old English)
Lover of the sea.

Marylou (33/6) (Hebrew-Old
German) Bitter battle maid.

Matilda (Old German) Powerful
in battle.

Melba (Latin) Mallow flower.

Melissa (Greek) Honeybee.

Melita (Hebrew) God's
vineyard.

Merna (Gaelic) Beloved.

Micheline (Hebrew) Like unto
the Lord.

Miki (Japanese) Three trees
together.

Millie (33/6) (Old German)
Industrious.

Minna (Old German) Love,
tender affection.

Mirabel (33/6) (Latin) Of
extraordinary beauty.

Miranda (33/6) (Latin)
Admirable.

Monserrat (Latin) Jagged
mountain.

Morgana (33/6) (Scottish
Gaelic) From the edge of the
sea.

Muriel (33/6) (Hebrew)
Bitterness.

Nalani (Hawaiian) Calmness of
the heavens.

Nancee (Hebrew) Full of grace.

Nari (Japanese) Thunderpeal.

Neala (Gaelic) The champion.

Neda (Old English) Rich and
happy protector.

Neiva (Spanish) Snow.

Nellwyn (33/6) (Greek) Friend
and companion.

Nerima (33/6) (Greek) Of the
sea.

Netta (Hebrew) Full of grace.

Ninette (33/6) (Spanish) Girl.

Nola (Latin) Symbol of peace.

Nyssa (Norse) Friendly elf or
brownie.

Odette (French) Home lover.

Oleen (Hawaiian) Joyous.

Olina (Hawaiian) Joyous.

Olwen (Welsh) White footprint.

Orah (Israeli) Light.

Oralie (33/6) (Latin) Golden.

Orelia (33/6) (Latin) Golden.

Palila (Polynesian) Bird.

Pamella (Greek) All honey.

Pamona (Latin) Fertile.

Pandora (33/6) (Greek) All-gifted.
Paula (Latin) Little.
Pauline (33/6) (Latin) Little.
Peggy (33/6) (Latin) A pearl.
Pelagia (33/6) (Greek) The sea.
Peta (Latin) Rock or stone.
Petra (Latin) Rock or stone.
Philippa (Greek) Lover of horses.
Phoebe (33/6) (Greek) Shining.
Phyllida (Greek) A green branch.
Pietra (33/6) (Latin) Rock or stone.
Pollyanne (Latin-Hebrew) Little, graceful.
Prue (Latin) Foresight.
Quintina (Latin) Fifth child.
Rae (Hebrew) Innocent as a lamb.
Rani (Sanskrit) Queen.
Rashida (33/6) (Arabic) Counselor.
Raven (Old English) Like the raven.
Raynell (33/6) (Sanskrit) Queen.
Rea (Greek) Flows from the earth.
Reeva (French) Riverside.
Renie (33/6) (Latin) Reborn.
Rhiamon (Welsh) A witch.
Rhonda (33/6) (Welsh) Grand.
Rihana (33/6) (Arabic) Sweet basil.

Rochelle (French) From the little rock.
Rocio (33/6) (Latin) Dew.
Rosamund (33/6) (Old German) Famous guardian.
Rosemary (English) Rose of Saint Mary.
Roza (Greek) A rose.
Ruella (Hebrew-Greek) Compassionate, light.
Rufina (33/6) (Latin) Red-haired.
Ruthann (33/6) (Hebrew) Compassionate, full of grace.
Sally (Hebrew) Princess.
Sami (Hebrew) His name is God.
Sascha (Greek) Helper of humankind.
Scarlet (Old English) Scarlet-colored.
Selene (Greek) Moon.
Selina (Greek) Moon.
Shani (African) Marvelous.
Sheenah (33/6) (Hebrew) God's gift of grace.
Sheryl (33/6) (Old English) From the bright meadow.
Shina (Japanese) Good, virtue.
Shirley (Old English) From the bright meadow.
Sidra (Latin) Related to the stars.
Sima (Hebrew) Treasure.
Soledad (Latin) Alone.

Stella (Latin) Star.
Suki (Hebrew) Graceful lily.
Suzu (Japanese) Long-lived.
Tallulah (American Indian)
    Leaping water.
Tamiko (Japanese) People.
Telma (Greek) A nursling.
Teodora (33/6) (Greek) Gift of
    God.
Thalia (Greek) Joyful or
    blooming.
Theodosia (Greek) Gift of God.
Tilly (Old German) Powerful in
    battle.
Tomasine (33/6) (Hebrew) The
    devoted sister.
Topaz (Latin) Topaz gem.
Tory (Latin) Victorious.
Tricia (33/6) (Latin) Wellborn
    noble.
Trista (Latin) Melancholy.
Trixy (33/6) (Latin) Bringer of
    joy.

Unique (33/6) (Latin) The only
    one.
Valora (Latin) Strong, valorous.
Verna (Latin) Springlike,
    youthful.
Veronica (Greek) Bringer of
    victory.
Viridiana (Latin) Green.
Vivianne (Latin) Alive.
Wilfreda (Old German)
    Resolute and peaceful.
Wilone (33/6) (Old English)
    Desired.
Wren (Old English) Wren.
Yesenia (33/6) (Arabic) Flower.
Yonina (33/6) (Hebrew) Dove.
Zalika (Swahili) Wellborn.
Zilla (Hebrew) Shadow.
Zoa (Greek) Life.
Zora (Greek) Dawn.
Zulema (Hebrew) Peace.

NUMBERS SIX AND THIRTY-THREE NAMES—MALE

Abbott (Hebrew) My father is
joy.

Achilles (33/6) (Greek) From
    the river Achillios.

Adair (Scottish Gaelic) From the oak tree ford.

Adwin (Ghana) Creative.

Aiden (Gaelic) Warmth of the home.

Allister (33/6) (Greek) Helper of humankind.

Allun (Gaelic) Handsome, cheerful.

Alvaro (Old German) All wise, prudent.

Andre (Greek) Strong and manly.

Ansel (Old French) Adherent of a nobleman.

Antoine (33/6) (Latin) Praiseworthy, without a peer.

Aram (Assyrian) High, exalted.

Arick (Old English) Sacred ruler.

Armand (Old German) Army man.

Asher (Hebrew) Divinely gifted.

Axel (Hebrew) Father of peace.

Barak (Hebrew) Flash of lightning.

Barric (33/6) (Hebrew) Flash of lightning.

Bartholomew (Hebrew) Son of a farmer.

Bayard (Old English) Having reddish-brown hair.

Beacher (33/6) (Old English) One who lives by oak trees.

Belden (Old English) Dweller in the beautiful glen.

Benny (Hebrew) Son of my right hand.

Benson (Old English) Son of Benjamin.

Berkley (33/6) (Old English) From the birch meadow.

Berle (Old English) Cup-bearer.

Beto (Old English) Noble and brilliant.

Bishop (33/6) (Old English) Bishop.

Blade (Old English) Knife, sword.

Boone (Old French) Good.

Booth (Old English) From the hut.

Borg (Old Norse) Castle dweller.

Bryan (Celtic) Strength with virtue and honor.

Burnard (33/6) (Old German) Brave as a bear.

Burne (Old English) Brook.

Cathmor (33/6) (Gaelic) Great warrior.

Cedric (33/6) (Old English) Battle chieftain.

Cesare (Latin) Long-haired, emperor.

Cheney (33/6) (Old French) Dweller in the oak forest.

Chester (33/6) (Old English) From the fortified camp.

Cheval (Old French) Horseman, knight.

Cleveland (33/6) (Old English) From the cliff-land.

Clinton (33/6) (Old English) From the hillside town.

Clive (Old English) From the cliff.

Corcoran (Gaelic) Reddish complexion.

Cordell (33/6) (Old French) Ropemaker.

Cortez (33/6) (Spanish) Courteous.

Coyle (Celtic) Leader in battle.

Cranley (33/6) (Old English) From the crane meadow.

Dagwood (33/6) (Old English) Forest of the shining one.

Daly (Gaelic) Assembly, gathering.

Damian (Greek) Constant.

Dane (Old English) From Denmark.

Darwin (33/6) (Old English) Beloved friend.

Deacon (Greek) Servant.

Dean (Old English) From the valley.

Delano (Old French) Of the night.

Demetrius (Greek) From the fertile land.

Dempsey (33/6) (Gaelic) Proud.

Denis (Greek) God of wine.

Denman (Old English) Resident in the valley.

Dion (Greek) God of wine.

Dirk (Old German) Leader of the people.

Dominick (Latin) Belonging to the Lord.

Don (Gaelic) World ruler.

Dougal (Celtic) From the dark stream.

Dow (Gaelic) Black-haired.

Druce (Celtic) Son of Drew.

Duarte (Old English) Prosperous guardian.

Dumont (Old French) From the mountain.

Durant (Latin) Enduring.

Dustin (Old German) Valiant fighter.

Duval (Old French) From the valley.

Edan (Celtic) Flame.

Eldredge (Old English) Old wise ruler.

Elford (33/6) (Old English) From the alder tree ford.

Ellsworth (Old English) Nobleman's estate.

Ely (Hebrew) Jehovah.

Emlyn (Latin) Flattering.

Enzo (Latin) Crowned with laurels.

Erasmus (Greek) Lovable.

Ernesto (33/6) (Old English) Earnest.

Erwin (33/6) (Old English) Sea friend.

Evan (Gaelic) Wellborn young warrior.

Evian (Hebrew) God's gracious gift.

Fabian (Latin) Bean grower.

Fabio (Latin) Bean grower.

Falcon (Latin) Falcon.

Felippe (Greek) Lover of horses.

Ferguson (Gaelic) The best choice.

Fernald (33/6) (Old English) From the fern slope.

Filmore (Old English) Very famous.

Fitzhugh (Old English) Son of the intelligent man.

Fonso (Old German) Noble and eager.

Fred (Old German) Peaceful ruler.

Freemont (Old German) Guardian of freedom.

Gabe (Hebrew) Devoted to God.

Galloway (33/6) (Celtic) Stranger from the Gaels.

Gamaliel (33/6) (Hebrew) God gives due reward.

Garry (33/6) (Old English) Spear-carrier.

Garvey (33/6) (Gaelic) Rough peace.

Garvyn (33/6) (Old German) Spear-friend ally.

Gary (Old English) Spear-carrier.

Gavyn (Old Welsh) White hawk.

Gayelord (Old French) Lively.

Gaylor (33/6) (Old French) Lively.

Genaro (33/6) (Spanish) Consecrated to God.

Geoffrey (Old German) God's divine peace.

Geovanni (Hebrew) God's gracious gift.

Geronimo (Latin) Sacred or holy name.

Gerson (33/6) (Hebrew) Alien resident.

Gilchrist (Gaelic) Servant of Christ.

Godfree (Old German) God's divine peace.

Goodwin (Old English) Friend of God.

Gordy (33/6) (Old English) Round hill.

Gowan (Gaelic) Wellborn young warrior.

Gradey (33/6) (Gaelic) Noble, industrious.

Grant (French) Great.

Gray (Old English) Son of the bailiff.

Griffin (Old Welsh) Fierce chief.

Gustavo (Swedish) Staff of the Goths.

Hakim (Arabic) Wise.

Hall (Old English) From the manor or hall.

Hans (Hebrew) God's gracious gift.

Harmon (33/6) (Latin) High-ranking person.

Heath (Old English) Heath or wasteland.

Heathcliff (Old English) From the heath cliff.

Hector (33/6) (Greek) Steadfast.

Herschel (Hebrew) Deer.

Hilton (33/6) (Old English) From the hill town.

Holbrook (Old English) From the brook in the hollow.

Howard (33/6) (Old English) Cliff guardian.

Howe (Old German) High.

Howie (33/6) (Old English) Cliff guardian.

Hugo (Old English) Intelligence.

Humbert (33/6) (Old German) Brilliant Hun.

Humphrey (Old German) Peaceful Hun.

Huntley (33/6) (Old English) Hunter's meadow.

Hymie (33/6) (Hebrew) Life.

Iain (Hebrew) God's gracious gift.

Ian (Hebrew) God's gracious gift.

Ibrahim (Hebrew) Father of the multitude.

Ichabod (33/6) (Hebrew) The glory has departed.

Isaac (Hebrew) He laughs.

Ivon (Old French) Archer with a yew bow.

Jaegar (Old German) Hunter.

Jasper (Persian) Treasurer.

Jedrick (33/6) (Polish) A strong man.

Jeremiah (Hebrew) Exalted by the Lord.

Jeriah (33/6) (Hebrew) Jehovah has seen.

Joaquin (33/6) (Hebrew) Jehovah has established.

Joel (Hebrew) Jehovah is God.

Jorgen (33/6) (Greek) Landholder, farmer.

Kalvin (Latin) Bald.

Kanoa (Hawaiian) The free one.

Karl (Old German) Strong and masculine.

Kennedy (33/6) (Gaelic) Helmeted chief.

Kenny (Gaelic) Handsome.

Kimball (Old Welsh) Warrior chief.

Kincaid (33/6) (Celtic) Battle chief.

Knight (33/6) (Old English) Mounted soldier.

Kolton (Old English) From the coal town.

Konnor (33/6) (Gaelic) Wise
man.

Korbin (33/6) (Latin) Raven.

Lachlan (Scottish Gaelic) From
the land of the lakes.

Landon (Old English) From the
long hill.

Latimer (33/6) (Old English)
Interpreter.

Leonard (33/6) (Old French)
Lion-brave.

Linford (Old English) From the
lime tree ford.

Livingston (Old English) From
Leif's town.

Lorenzo (Latin) Crowned with
laurels.

Lucian (Latin) Crowned with
laurels.

Ludlow (Old English) From the
prince's hill.

Malcolm (Scottish Gaelic)
Disciple of Saint Columbia.

Malek (Arabic) Master.

Manzo (Japanese) Third son.

Marmaduke (33/6) (Celtic) Sea
leader.

Martel (Old French) Mace.

Maximilian (Latin) Most
excellent.

Melbourne (Old English) From
the mill stream.

Merlyn (33/6) (Old English)
Falcon or hawk.

Merrill (Old French) Famous.

Millard (33/6) (Latin) Caretaker
of the mill.

Miller (33/6) (Latin) Caretaker
of the mill.

Minor (33/6) (Latin) Junior,
younger.

Moe (Latin) Dark-skinned.

Montague (33/6) (Latin) From
the pointed mountain.

Monty (Latin) From the pointed
mountain.

Morven (33/6) (Gaelic) Blond
giant.

Moshe (Hebrew) Drawn out of
the water.

Murray (33/6) (Scottish Gaelic)
Sailor.

Noland (Gaelic) Famous or
noble.

Norm (Old French) A northman.

Norton (33/6) (Old English)
From the northern town.

Oakes (Old English) From the
oak trees.

Oakley (Old English) From the
oak tree field.

Oates (Greek) Keen of hearing.

Obadias (Hebrew) Servant of
God.

Olivero (Latin) Symbol of
peace.

Oliviero (Latin) Symbol of
peace.

Orlan (Old English) From the
pointed land.

Orvin (33/6) (Old English) Spear-friend.

Ossy (Old English) Divine spearman.

Palladin (33/6) (American Indian) Fighter.

Parke (Old English) Guardian of the park.

Parker (33/6) (Old English) Guardian of the park.

Parkin (33/6) (Old English) Little Peter.

Parnell (33/6) (Old French) Little Peter.

Parrnell (Old French) Little Peter.

Parry (33/6) (Old English) From the churchyard.

Patrick (33/6) (Latin) Wellborn, noble.

Pepin (33/6) (Old German) Petitioner.

Philo (33/6) (Greek) Loving, friendly.

Placido (33/6) (Latin) Peaceful, serene.

Price (33/6) (Old Welsh) Son of the ardent one.

Proctor (Latin) Administrator.

Purcell (33/6) (Old French) Valley-piercer.

Purvis (33/6) (English-French) To provide food.

Radburn (33/6) (Old English) · From the red stream.

Raleigh (Old English) From the deer meadow.

Rashad (Arabic) Counselor.

Rasheed (33/6) (Arabic) Counselor.

Rawlins (33/6) (Anglo-French) Son of little counsel-wolf.

Raymund (33/6) (Old German) Mighty or wise protector.

Reade (Old English) Red-haired.

Renard (33/6) (Old French) Fox.

Rhodes (33/6) (Old English) Dweller at the crucifixes.

Rigel (33/6) (Arabic) Foot.

Rinji (33/6) (Japanese) Peaceful forest.

Robert (33/6) (Old English) Bright fame.

Rodin (33/6) (Old English) From the reed valley.   .

Rolf (Old English) Swift wolf.

Rolph (33/6) (Old English) Swift wolf.

Rowland (33/6) (Old German) From the famous land.

Ruben (Hebrew) Behold a son.

Sam (Hebrew) His name is God.

Sancho (Latin) Sanctified.

Sansone (Hebrew) Like the sun.

Sayers (Welsh) Carpenter.

Sayres (Welsh) Carpenter.

Seamus (Hebrew) The supplanter.

Seldon (Old English) From the willow tree valley.

Shaw (Old English) From the grove.

Sheehan (33/6) (Gaelic) Little and peaceful.

Sherman (33/6) (Old English) Wool-shearer.

Sherwin (Old English) Swift runner.

Sigmund (33/6) (Old German) Victorious protector.

Silas (Latin) Forest god.

Simpson (33/6) (Hebrew) Like the sun.

Skerry (33/6) (Scandinavian) From the rocky island.

Skippy (33/6) (Scandinavian) Shipmaster.

Smith (Old English) Blacksmith.

Somerset (33/6) (Old English) Place of the summer settlers.

Sonny (English) Son.

Sorrel (33/6) (Old French) Reddish-brown.

Spense (Old English) Dispenser of provisions.

Sprague (33/6) (French) Lively.

Stamford (33/6) (Old English) From the stormy crossing.

Stanley (Old English) From the rocky meadow.

Stephen (33/6) (Greek) Crowned.

Stillmann (33/6) (Old English) Quiet man.

Sven (Scandinavian) Youth.

Sweeney (33/6) (Gaelic) Little hero.

Swinton (33/6) (Old English) From the pig farm.

Talbert (Old German-French) Valley, bright.

Tanjiro (33/6) (Japanese) High-valued second son.

Tau (African) Lion.

Thaxter (33/6) (Old English) Roof thatcher.

Thorald (33/6) (Scandinavian) Thor's ruler.

Tim (Greek) Honoring God.

Timmie (33/6) (Greek) Honoring God.

Townsend (33/6) (Old English) From the town's end.

Troy (Gaelic) Foot soldier.

Truman (Old English) Faithful man.

Tucker (Old English) Tucker of cloth.

Turner (33/6) (Latin) Lathe worker.

Urbanus (Latin) From the city.

Urson (Latin) Little bear.

Verdell (33/6) (Latin) Green, flourishing.

Victor (33/6) (Latin) Victorious.

Vincent (33/6) (Latin) Victorious.

Wade (Old English) Advancer.
Wainwright (Old English) Wagon-maker.
Waldon (Old English) From the woods.
Waldron (33/6) (Old German) Strength of the raven.
Watford (33/6) (Old English) From the hurdle by the ford.
Weston (Old English) From the western estate.
Whitby (33/6) (Old English) From the white field.
Wilber (33/6) (German-French) Brilliant hostage.
Wilburt (33/6) (German-French) Brilliant hostage.
Wilkie (33/6) (Old German) Determined guardian.

Winn (Old English) From the friendly place.
Winston (33/6) (Old English) From the friend's town.
Winthrop (Old English) Dweller at the friend's estate.
Woolsey (33/6) (Old German) Victorious wolf.
Ximenez (Hebrew) One who hears.
York (Celtic) Yew tree estate.
Yuma (American Indian) Son of a chief.
Zeb (Hebrew) The Lord's gift.
Zedekiah (Hebrew) God is mighty and just.
Zeno (Greek) Life.

NUMBERS SIX AND THIRTY-THREE NAMES—UNISEX

Alex (Greek) Helper of humankind.
Alexi (Greek) Helper of humankind.
Arden (Latin) Ardent, fiery.
Arnett (Old German) Little eagle.

Billy (Old German) Determined guardian.
Blair (Gaelic) From the plain.
Britt (Latin) From England.
Cameron (33/6) (Scottish Gaelic) Crooked nose.
Cass (Gaelic) Clever.

Connie (33/6) (Latin) Firmness, constancy.

Dagny (Scandinavian) Day, brightness.

Darcy (Old French) From the fortress.

Daren (Old French) From the fortress.

Darren (33/6) (Gaelic) Great.

Darryl (33/6) (Old French) Beloved.

Daryl (Old French) Beloved.

Dennie (33/6) (Greek) God of wine.

Diamond (33/6) (Greek) Of high value, brilliant.

Dore (Greek) Gift.

Flo (Latin) Flowering or blooming.

Florence (Latin) Flowering or blooming.

Freddie (Old German) Peaceful ruler.

Hailey (33/6) (Old English) Hay meadow.

Haley (Old English) Hay meadow.

Harley (33/6) (Old English) From the long field.

Jessy (Hebrew) Wealthy.

Kim (Old English) From the king's wood.

Kory (Gaelic) From the hollow.

Lesley (Scottish Gaelic) From the grey fortress.

Lonnie (33/6) (Old English) Solitary.

Lyn (Old English) From the waterfalls.

Mackenzie (Gaelic) Son of the wise leader.

Macy (Old French) Mace.

Mallory (33/6) (Old German) Army counselor.

Marlowe (33/6) (Old English) From the hill by the lake.

Michael (33/6) (Hebrew) Like unto the Lord.

Montana (Latin) Mountain.

Nickie (33/6) (Greek) The people's victory.

Noele (French) Born at Christmas.

Palm (Latin) The palm-bearer.

Rande (Old English) Swift wolf.

Rebel (Latin) Opposer, revolutionary.

Reggie (Old English) Mighty and powerful.

Rene (Latin) Reborn.

Riley (33/6) (Gaelic) Valiant.

Robbie (33/6) (Old English) Bright fame.

Rodie (33/6) (Latin) Dew of the sea.

Ryun (Gaelic) Little king.

Shea (Gaelic) From the fairy fort.

Sheridan (Celtic) Wild man or savage.

Skye (Dutch) Sheltering.
Taryn (Gaelic) Rocky pinnacle.
Teddi (Greek) Gift of God.
Tierney (Gaelic) Lordly one.
Tobe (Hebrew) The Lord is
  good.
Vann (Dutch) Of noble descent.
Weslee (Old English) From the
  western meadow.

11

number Seven

Qualities of number Seven

Keywords: mystical, philosophical, contemplative, artistic, imaginative, creative, charitable, intuitive, intellectual, spiritual, wise, silent, secretive, solitary, psychic, refined, probing, independent, religious, depressive, moody, aloof, lazy, impractical, confused, superficial, crafty, cunning, deceitful, withdrawn, phobic.

Symbolism: The number Seven represents mysticism. It is an indivisible number and stands for the "God force" in nature. Traditionally God rested on the seventh day; therefore all things rest, reflect, and contemplate under seven's vibration. Masculine in nature, seven is symbolized by Neptune and the seven-branched candelabra. Its element is water and it is related to the days of the week, the seven chakras, the lotus, the ziggurat (seven-stepped ladder to heaven), and the Masonic apron, which is composed of the triangle above the square.

Personality: The number seven personality is creative, imaginative, and likely to be deeply interested in all aspects of mysticism and the occult. Sevens have deep reservoirs of wisdom, but often have difficulty bringing it to the surface.

Many Sevens are clairvoyant or gifted with ESP. Musical, poetic, and artistic ability is quite common among Sevens, and they like to indulge in handicrafts, writing, dancing, and painting. Seven is a spiritual number, and Sevens tend to be deeply religious in a most unorthodox fashion. Occasionally they create religions of their own based on mystical beliefs. Sevens are always a mystery to others, and their solitary nature is much given to contemplation and meditation. They make exceptionally good counselors as they are able to think in the abstract, visualize the future, and allow their spirits to reach deep into the unknown areas where truths are revealed. Sevens usually work best in an individual capacity where they can be the absolute authorities within their own realms.

Sevens can be depressive or moody and must take care to separate imaginary fears and phobias from the realities of life. Most Sevens need a push to get going, but once they start they will usually see things through to the end. Because they breathe the rarefied air of the visionary, they are seldom understood or appreciated, and this can lead to frustration and disappointment for them.

Careers: accountant, actor, analyst, artist, astronomer, astrologer, healer, historian, minister, musician, navigator nurse, occultist, philosopher, photographer, poet, psychiatrist, psychologist, researcher, sailor, secret agent, scientist, tarot card reader, technician, underwater explorer, writer.

NUMBER SEVEN NAMES—FEMALE

Adara (Arabic) Virgin.
Adelita (Old German) Of noble rank.
Adena (Hebrew) Sensuous.
Adrienne (Latin) Black earth.
Alanna (Gaelic) Comely, cheerful.
Alethea (Greek) Truthful.
Alexa (Greek) Helper of humankind.
Alexia (Greek) Helper of humankind.
Alika (Greek) Helper of humankind.
Alison (Greek) Truthful.
Alissa (Greek) Truthful.
Alita (Old German) Of noble rank.
Aloise (Old German) Famous in war.
Alta (Latin) High or lofty.
Alyda (Greek) Beautifully dressed.
Amabel (Latin) Lovable.
Amanda (Latin) Worthy of being loved.
Amara (Greek) Of eternal beauty.
Ameera (Arabic) Princess.
Ana (Hebrew) Full of grace.

Anabelle (Hebrew-Latin) Graceful, beautiful.
Analise (Hebrew) Graced with God's promise.
Andrea (Latin) Womanly.
Angelica (Greek) Heavenly messenger.
Anne (Hebrew) Full of grace.
Annelise (Hebrew) Graced with God's promise.
Annette (Hebrew) Full of grace.
Annie (Hebrew) Full of grace.
Arabella (Latin) Beautiful altar.
Ariadne (Greek) Holy one.
Armanda (Old German) Army woman.
Atalanta (Greek) Mighty adversary.
Baptista (Latin) Free from sin.
Barbara (Latin) Beautiful stranger.
Beatrix (Latin) Bringer of joy.
Benicia (Latin) Blessed.
Benigna (Latin) A great lady.
Berenice (Greek) Bringer of victory.
Bessy (Hebrew) God's promise.
Bette (Hebrew) God's promise.
Binah (Hebrew) Understanding.
Bliss (Old English) Joy, bliss.

Blondie (French) Little fair one.

Bonita (Spanish) Pretty.

Bonny (Scottish-English) Beautiful, pretty.

Branwyn (Old Welsh) White raven.

Brier (Old English) A thorny plant.

Callan (Gaelic) Powerful in battle.

Caresse (French) Beloved.

Carissa (Greek) Loving.

Carita (Latin) Charity.

Carolyn (Old German) Little woman born to command.

Casandra (Greek) Disbelieved by men.

Cassia (Greek) Sweet-scented spice.

Catharine (Greek) Pure maiden.

Cedrica (Old English) Battle chieftain.

Chanel (French) Channel.

Chelsie (Old English) A port of ships.

Cher (Old French) Beloved.

Chiquita (Spanish) Little one.

Chloe (Greek) Young grass.

Christabel (Greek) Beautiful anointed one.

Claribelle (French) Brilliant, beautiful.

Corinthia (Greek) Woman of Corinth.

Cosina (Greek) World harmony.

Cybele (Greek) A prophetess.

Deana (Latin) Divine moon goddess.

Deeann (Latin) Divine moon goddess.

Delcine (Latin) Sweet and charming.

Delicia (Latin) Gives pleasure.

Della (Greek) Visible.

Demeter (Greek) From the fertile land.

Devona (Old English) Defender of Devonshire.

Devonne (Old English) Defender of Devonshire.

Dianna (Latin) Divine moon goddess.

Dianthe (Latin-Greek) Divine moon goddess.

Dinora (Hebrew) Judged.

Dionne (Greek) Daughter of heaven and earth.

Dolores (Spanish) Sorrows.

Doreen (Gaelic) Sullen.

Ebony (Greek) A hard dark wood.

Edana (Celtic) Flame.

Eirlys (Old Welsh) Snow drop.

Eleanor (Old French) Light.

Elga (Slavic) Consecrated.

Elizabeth (Hebrew) God's promise.

Ellie (Greek) Light.

Eloisa (Old German) Famous woman warrior.

Ethyl (Old English) Noble.
Eulalia (Greek) Sweet-spoken.
Eustacia (Greek) Steadfast.
Evania (Greek) Tranquil.
Farah (Old English) Beautiful, pleasant.
Farrah (Old English) Beautiful, pleasant.
Fern (Old English) A fern.
Fiammetta (Italian) A flickering fire.
Flora (Latin) Flowering or blooming.
Florinda (Latin) Flowering or blooming.
Flower (Latin) Flower.
Francesca (Latin) Free.
Francine (Latin) Free.
Freda (Old German) Peaceful ruler.
Frieda (Old German) Peaceful ruler.
Fritzi (Old German) Peaceful ruler.
Gem (Latin) Precious stone.
Geovanna (Hebrew) God's gracious gift.
Gitana (Old English) Wanderer.
Glenda (Old Welsh) Dweller in a valley.
Goldie (Old English) Golden.
Grace (Latin) Graceful.
Gracie (Latin) Graceful.
Guinevere (Old Welsh) White-browed.

Gwynne (Old Welsh) White-browed.
Hadassah (Hebrew) Myrtle tree.
Harmonia (Latin) Harmony.
Harriet (French) Mistress of the household.
Hasina (Swahili) Good.
Hazel (English) Hazelnut tree.
Hilda (Old German) Battle maid.
Honor (Latin) Honorable.
Huette (Old English) Brilliant thinker.
Hyacinth (Greek) Hyacinth flower.
Hynda (Hebrew) Female deer.
Idella (Celtic) Bountiful.
Ingrid (Norse) Hero's daughter.
Iolana (Greek) Violet-colored dawn.
Ione (Greek) Violet-colored stone.
Isabella (Old Spanish) Consecrated to God.
Italia (Latin) From Italy.
Ivey (Old English) Ivy vine.
Ivonne (Old French) Archer with a yew bow.
Jacqueline (Hebrew) The supplanter.
Jenevieve (French) Pure white wave.
Jill (Greek) Young in heart and mind.

Jina (Latin) Queen.

Joaquina (Hebrew) Jehovah has established.

Joby (Hebrew) The afflicted.

Joleen (Hebrew) He shall add.

Jolene (Hebrew) He shall add.

Kaila (Israeli) The laurel crown.

Kala (Hindi) Black, time.

Kalinda (Sanskrit) Sun.

Kami (Japanese) Lord.

Kanya (Hindi) Virgin.

Karla (Old German) Little woman born to command.

Kassandra (Greek) Disbelieved by men.

Katelyn (Greek) Pure maiden.

Kathryn (Greek) Pure maiden.

Keilani (Hawaiian) Glorious chief.

Kelsie (Old Norse) From the ship island.

Ketifa (Arabic) Flowering.

Kima (Japanese) Happiness.

Kioko (Japanese) Child born with happiness.

Kirima (Eskimo) A hill.

Klara (Latin) Bright, shining girl.

Krystal (Latin) Without deception.

Ladonna (French) The lady.

Larissa (Greek) Cheerful.

Latisha (Latin) Gladness.

Lila (Latin) Lily flower.

Lilith (Arabic) Of the night.

Liv (Latin) Symbol of peace.

Lizette (Hebrew) God's promise.

Lorelle (Latin) Little.

Lucianne (Latin) Crowned with laurels.

Lucilla (Latin) Light.

Lucy (Latin) Light.

Lumina (Latin) Brilliant, illuminated.

Lupita (Arabic) River of black stones.

Lura (German) Siren.

Lynne (Old English) From the waterfalls.

Lynnette (Old French) Linnet bird.

Lysette (Hebrew) God's promise.

Magali (Hebrew) Woman of Magdala.

Mandisa (African) Sweetness.

Margarita (Latin) A pearl.

Mariann (Hebrew) Bitter, graceful.

Marigold (English) Mary's gold.

Marisa (Latin) Of the sea.

Martha (Arabic) Lady, mistress.

Meg (Latin) A woman without equal.

Merry (English) Mirthful, joyous.

Michaela (Hebrew) Like unto the Lord.

Millicent (Old German) Industrious.

Mirella (Latin) Wonderful.

Mona (Greek) Solitary.

Mozelle (Hebrew) Drawn out of the water.

Nadeen (Russian) Hope.

Nanette (Hebrew) Full of grace.

Naomi (Hebrew) Pleasant.

Nara (American Indian) Oak.

Nasaya (Hebrew) Miracle of God.

Nathalie (Latin) Natal day.

Nazneen (Persian) Exquisitely beautiful, charming.

Nell (Old French) Light.

Noelani (Hawaiian) Beautiful one from heaven.

Noellyn (French) Born at Christmas.

Norma (Latin) A rule, pattern, or precept.

Nova (Latin) New.

Nyla (Gaelic) the champion.

Oletha (Scandinavian) Light, nimble.

Oliana (Polynesian) Oleander.

Ondine (Latin) Wave, of water.

Onida (American Indian) The looked-for one.

Ophelie (Greek) Serpent.

Oria (Israeli) Light.

Palma (Latin) The palm-bearer.

Palmira (Latin) The palm-bearer.

Paz (Latin) Peace.

Pearl (Latin) A pearl.

Penelope (Greek) The weaver.

Perla (Latin) A pearl.

Phedra (Greek) Bright.

Philana (Greek) Lover of mankind.

Poppy (Latin) Poppy flower.

Portia (Latin) Offering.

Precious (Old French) Of great value.

Quintana (Latin) Fifth child.

Quintilla (Latin) Fifth child.

Radmilla (Slavic) Worker for the people.

Rana (Sanskrit) Queen.

Randene (Old English) Swift wolf.

Rasheeda (Arabic) Counselor.

Rebeca (Hebrew) The captivator.

Reena (Hebrew) Song.

Remy (French) From Rheims.

Renae (Latin) Reborn.

Rheba (Hebrew) The captivator.

Rheta (African) To shake.

Rhianon (Welsh) A witch.

Rhody (Latin) Dew of the sea.

Ricca (Old German) Peaceful ruler.

Roberta (Old English) Bright fame.

Rodina (Old English) From the reed valley.

Romina (Latin) Woman of Rome.

Ronda (Welsh) Grand.

Rosalie (Latin) Rose.

Rosalinde (Spanish) Beautiful rose.

Rula (Latin) Sovereign ruler.

Sadira (Persian) Lotus.

Salena (Latin) Salty.

Santana (Spanish) Saint Anne.

Sapphira (Greek) Sapphire stone.

Seema (Hebrew) Treasure.

Serenity (Latin) Calm, serene.

Shaina (Hebrew) Beautiful.

Shana (Gaelic) Small and wise.

Shantel (Latin) Song.

Sharmaine (Latin) Song.

Sheena (Hebrew) God's gift of grace.

Shifra (Hebrew) Beautiful.

Sian (Hebrew) God's gracious gift.

Sierra (Latin) Mountainous.

Sinead (Hebrew) God's gracious gift.

Stefanie (Greek) Crowned.

Stephanie (Greek) Crowned.

Sumiko (Japanese) Child of goodness.

Susannah (Hebrew) Graceful lily.

Sylvia (Latin) From the forest.

Tabitha (Greek) Gazelle.

Talia (Greek) Joyful or blooming.

Tami (Japanese) People.

Tansy (Greek) Immortality.

Tanya (Greek) Giant.

Teresita (Greek) The harvester.

Tessy (Greek) The harvester.

Thea (Greek) Goddess.

Theola (Greek) Gift of God.

Thera (Greek) Wild, untamed.

Tiffanie (Greek) Appearance of God.

Trinidad (Latin) Triad.

Trinity (Latin) Triad.

Trudy (Old German) Beloved.

Ula (Celtic) Sea jewel.

Ursulina (Latin) Little bear.

Vanna (Greek) Butterflies.

Vicky (Latin) Victorious.

Victoria (Latin) Victorious.

Vita (Latin) Alive.

Wanda (Old German) The wanderer.

Wilhelmina (Old German) Determined guardian.

Winifred (Old German) Peaceful friend.

Winna (African) Friend.

Xuxa (Hebrew) Graceful lily.

Ynez (Greek) Pure, chaste.

Yvette (Old French) Archer with a yew bow.

NUMBER SEVEN NAMES—MALE

Abdullah (Arabic) Servant of God.

Abelard (Old German) Noble and resolute.

Adon (Hebrew) Lord God.

Alben (Latin) White.

Albern (Old English) Noble warrior.

Alfredo (Old English) Good counselor.

Alger (Old German) Noble spearman.

Altman (Old German) Wise old man.

Alyn (Gaelic) Handsome, cheerful.

Amon (Hebrew) Trustworthy, faithful.

Anders (Greek) Strong and manly.

Andres (Greek) Strong and manly.

Anselmo (Old German) Divine warrior.

Anthony (Latin) Praiseworthy, without a peer.

Antonio (Latin) Praiseworthy, without a peer.

Armon (Hebrew) High place.

Arnoldo (Old German) Strong as an eagle.

Arsenius (Greek) Potent.

Artemus (Greek) Gift of Artemis.

Ase (Hebrew) Physician.

Audel (Old English) Old friend.

Aurelius (Latin) Golden.

Averill (Latin) Forthcoming.

Baird (Gaelic) Ballad singer.

Bancroft (Old English) From the bean field.

Bard (Gaelic) Poet and singer.

Barton (Old English) From the barley farm.

Basil (Latin) Kingly, magnificent.

Baxter (Old English) Baker.

Beale (Old French) Handsome.

Bellamy (Old French) Handsome friend.

Benton (Old English) From the bent grass town.

Boyne (Gaelic) A rare person.

Brad (Old English) From the broad meadow.

Bram (Gaelic) Raven.

Brenton (Old English) From the town on the steep hill.

Brick (Old English) Bridge.

Bronson (Old English) Son of the dark-skinned one.

Bruno (Old German) Brown-haired.

Buell (Old German) Hill dweller.

Burt (Old English) From the fortress.

Cal (Latin) Bald.

Calder (Old English) The stream.

Calvin (Latin) Bald.

Carl (Old German) Strong and masculine.

Carleton (Old English) Farmer's town.

Carlisle (Old English) From the fortified town.

Carson (Old English) Son of the marsh dwellers.

Chad (Old English) Warlike.

Chaim (Hebrew) Life.

Chalmers (Scottish Gaelic) Son of the overseer.

Chance (English) Fortune, a gamble.

Che (Hebrew) He shall add.

Cherokee (American Indian) People of a different speech.

Chesney (Old French) Dweller in the oak forest.

Chick (Old German) Strong and masculine.

Clarence (Latin) Bright, famous.

Clifton (Old English) From the cliff-town.

Colm (Latin) Dove.

Colton (Old English) From the coal town.

Connor (Gaelic) Wise man.

Constant (Latin) Firmness, constancy.

Corbin (Latin) Raven.

Cornell (Latin) Yellow, horn-colored.

Crawford (Old English) From the ford of the crow.

Crispin (Latin) Curly-haired.

Cutler (Old English) Knifemaker.

Damario (Greek) Gentle.

Darrell (Old French) Beloved.

Davy (Hebrew) Beloved.

Delroy (Old French) Of the king.

Delton (Old English) From the town in the valley.

Denzil (Greek) God of wine.

Derek (Old German) Leader of the people.

Derrek (Old German) Leader of the people.

Derry (Gaelic) Great.

Dinsmore (Gaelic) From the fortified hill.

Douglas (Scottish Gaelic) From the dark water.

Doyle (Gaelic) Dark stranger.

Dryden (Old English) From the dry valley.

Dunham (Celtic) Dark man.

Edmund (Old English) Prosperous protector.

Elgar (Old German) Sword.

Ellard (Old German) Nobly brave.

Engelbert (Old German) Bright as an angel.

Ennis (Gaelic) The only choice.

Ephraim (Hebrew) Very fruitful.

Erik (Old Norse) Ever-powerful, ever-ruler.

Esmond (Old English) Gracious protector.

Estefan (Greek) Crowned.

Ewan (Gaelic) Wellborn young warrior.

Filip (Greek) Lover of horses.

Finn (Gaelic) Fair.

Finnegan (Gaelic) Fair.

Fitz (Old English) Son of.

Flint (Old English) Stream.

Ford (Old English) River crossing.

Fortino (Latin) Fortunate.

Fowler (Old English) Trapper of wild fowl.

Francisco (Latin) Free.

Frederick (Old German) Peaceful ruler.

Fritz (Old German) Peaceful ruler.

Fulton (Old English) A field near the town.

Gabian (Hebrew) Devoted to God.

Gabor (Hebrew) Devoted to God.

Gamal (Arabic) Camel.

Gates (Old English) Gatekeeper.

Georg (Greek) Landholder, farmer.

Gerhard (Old English) Spear-strong.

Gilberto (Old German) Brilliant hostage.

Giles (Greek) Shield-bearer.

Gilmore (Gaelic) Devoted to the Virgin Mary.

Girvin (Gaelic) Little rough one.

Giuliano (Greek) Young in heart and mind.

Gladwin (Old English) Cheerful.

Grange (Old English) Farmer.

Granger (Old English) Farmer.

Gregor (Latin) Watchman.

Gunner (German) Battler, warrior.

Guthrie (Gaelic) From the windy place.

Hakeem (Arabic) Wise.

Halim (Arabic) Gentle.

Halsey (Old English) From Hal's island.

Halstead (Old English) From the manor.

Hank (Old German) Lord of the manor.

Harry (Old German) Lord of the manor.

Harte (Old English) Male deer.

Harvey (Old German) Army warrior.

Hastings (Old English) Son of the stern man.

Hawk (English) A hawk, falcon.

Henry (Old German) Lord of the manor.

Hernando (Old German) World-daring.

Huntington (Old English) Hunting estate.

Huw (Old German) Brilliant thinker.

Hyman (Hebrew) Life.

Iaian (Hebrew) God's gracious gift.

Immanuel (Hebrew) God is with us.

Irving (Old English) Sea friend.

Isidore (Greek) Gift of Isis.

Jack (Hebrew) The supplanter.

Jarad (Hebrew) One who rules.

Jarvis (Old German) Keen with a spear.

Jimmy (Hebrew) The supplanter.

Jiro (Japanese) Second son.

Joed (Hebrew) Jehovah is witness.

Jomei (Japanese) Spread life.

Josh (Hebrew) Jehovah saves.

Jove (Latin) Father of the sky.

Kadir (Arabic) Spring greening.

Kallum (Gaelic) Mild, gentle.

Karim (Arabic) Noble, exalted.

Karsten (Greek) Anointed.

Kearney (Gaelic) Victorious.

Kedrick (Old English) Gift of splendor.

Keegan (Gaelic) Little and fiery.

Keir (Celtic) Dark-skinned.

Keitaro (Japanese) Blessed.

Kenton (Old English) From the king's estate.

Kerr (Old Norse) Marshland.

Kevin (Gaelic) Gentle, kind, and lovable.

Kurt (Old German) Bold counselor.

Laurence (Latin) Crowned with laurels.

Lawford (Old English) From the ford on the hill.

Lenny (Old French) Lion-brave.

Leopold (Old German) Bold for the people.

Lester (Latin) From the chosen camp.

Lincoln (Old English) From the settlement by the pool.

Lowell (Old French) Little wolf.

Luis (Old German) Famous warrior.

Maddox (Welsh) Beneficent.

Manfred (Old English) Man of peace.

Mano (Hawaiian) Shark.

Manolo (Hebrew) God is with us.

Marcel (Latin) Follower of Mars.

Mareo (Japanese) Rare, uncommon.

Marino (Latin) From the sea.

Mark (Latin) Follower of Mars.

Mathew (Hebrew) Gift of God.

Maurice (Latin) Dark-skinned.

Mervyn (Old English) Lover of the sea.

Mitchel (Hebrew) Like unto the Lord.

Morio (Japanese) Forest boy.

Morley (Old English) From the meadow on the moor.

Morly (Old English) From the meadow on the moor.

Morse (Old English) Son of the dark-complexioned one.

Muir (Scottish Gaelic) Moor.

Myer (Hebrew) Bringer of light.

Nardo (Latin) Strong, hardy.

Nelson (English) Neil's son.

Nero (Latin) Stern.

Neville (Latin) From the new town.

Newman (Old English) New arrival.

Noe (Hebrew) Rest, comfort.

Normand (Old French) A northman.

Norrie (Old French) Northerner.

Northrop (Old English) From the north farm.

Olaf (Old Norse) Ancestral talisman.

Oren (Gaelic) Pale-complected.

Orlando (Old English) From the pointed land.

Orman (Old German) Mariner or shipman.

Ormond (Old English) Spear-protector.

Orren (Gaelic) Pale-complected.

Osborne (Old English) Warrior of God.

Otto (Old German) Rich.

Pascal (Latin) Born at Easter or Passover.

Paxon (Latin) From the peaceful town.

Payne (Latin) Villager.

Philip (Greek) Lover of horses.

Pinchas (Hebrew) Oracle.

Porfirio (Greek) Dressed in purple.

Quinlan (Gaelic) Wise, intelligent.

Raddy (Old English) From the red stream.

Rafael (Hebrew) Healed by God.

Rafaello (Hebrew) Healed by God.

Rafi (Arabic) Exalting.

Ramon (Old German) Mighty or wise protector.

Randolf (Old English) Shield-wolf.

Randolph (Old English) Shield-wolf.

Raphael (Hebrew) Healed by God.

Raul (Old English) Swift wolf.

Redford (Old English) Red river crossing.

Redmund (Old German) Protecting counselor.

Reginald (Old English) Mighty and powerful.

Remington (Old English) From the raven estate.

Rhys (Welsh) Enthusiastic.

Richard (Old German) Wealthy, powerful.

Rigby (Old English) Valley of the rulers.

Riordan (Gaelic) Bard, royal poet.

Rip (Old English) From the shouter's meadow.

Robinson (English) Son of Robin.

Rolando (Old German) From the famous land.

Romain (Latin) From Rome.

Roman (Latin) From Rome.

Rosco (Old Norse) From the deer forest.

Roth (Old German) Red hair.

Rourke (Gaelic) Famous ruler.

Russell (Old French) Red-haired.

Rutherford (Old English) From the cattle ford.

Ryder (Old English) Horseman.

Sampson (Hebrew) Like the sun.

Sander (Old English) Son of Alexander.

Santos (Latin) Saints.

Seiji (Japanese) Lawful.

Selig (Old German) Blessed.

Sergent (Old French) Army officer.

Seth (Hebrew) Appointed, substitute.

Seward (Old English) Sea guardian.

Sexton (Old English) Church official.

Shaan (Hebrew) Peaceful.

Sharif (Arabic) Illustrious.

Simon (Hebrew) One who hears.

Smitty (Old English) Blacksmith.

Sparke (Old English) Happy.

Stanford (Old English) From the rocky ford.

Stewart (Old English) The steward.

Swen (Scandinavian) Youth.

Tabib (Turkish) Physician.

Tad (Greek) Stout-hearted.

Takeo (Japanese) Strong like bamboo.

Talbot (Old German-French) Valley, bright.

Tally (Old German-French) Valley, bright.

Tavish (Gaelic) Twin.

Terence (Latin) Smooth.

Terrence (Latin) Smooth.

Theodoric (Old German) Ruler of the people.

Thor (Old Norse) Thunder.

Thornton (Old English) From the thorny farm.

Todd (Old English) A fox.

Trenton (Old English) Town on the river Trent.

Trever (Gaelic) Wise and discreet.

Trevin (Welsh) Fair town.

Tripp (Old English) To dance or hop.

Truemann (Old English) Faithful man.

Tyrone (Gaelic) Land of Owen.

Ugo (Nigerian) Eagle.

Valentin (Latin) Strong.

Vernon (Latin) Springlike, youthful.

Vladimir (Old Slavic) Powerful prince.

Walker (Old English) Thickener of cloth.

Walter (Old German) Powerful warrior.

Warner (Old German) Defending army or warrior.

Warren (Old German) Watchman or defender.

Waverly (Old English) Quaking aspen tree meadow.

Whitman (Old English) White-haired man.

Whittaker (Old English) From the white field.

Willard (Old German) Resolutely brave.

William (Old German) Determined guardian.

Wilmus (Old German) Determined guardian.

Winslow (Old English) From the friend's hill.

Wolfie (Old German) Wolf.

Wolfram (Old German) Respected and feared.

Wyndham (Scottish Gaelic) Village near the winding road.

Xaver (Arabic) Bright.

Xavier (Arabic) Bright.

Yale (Old English) From the corner of the land.

Yates (Old English) Dweller at the gates.

Zarek (Slavic) God protects.

Zephaniah (Hebrew) Treasured by the Lord.

NUMBER SEVEN NAMES—UNISEX

Alexis (Greek) Helper of humankind.
Ardell (Latin) Ardent, fiery.
Ashley (Old English) From the ash tree meadow.
Bethel (Hebrew) House of God.
Blaine (Gaelic) Thin, lean.
Blane (Gaelic) Thin, lean.
Brook (Old English) From the brook.
Carey (Old German) Strong and masculine.
Carmel (Hebrew) God's vineyard.
Carroll (Old German) Noble and strong.
Channing (Old English) Knowing.
Cory (Gaelic) From the hollow.
Dakota (American Indian) Friend, ally.
Daron (Gaelic) Great.
Doran (Celtic) Stands firm.
Dorian (Greek) From the sea.
Dru (Latin) Strong.
Francis (Latin) Free.
Gael (Old English) Gay, lively.
Gale (Old English) Gay, lively.
Glenn (Old Welsh) Dweller in a valley.

Guadalupe (Arabic) River of black stones.
Jan (Hebrew) God's gift of grace.
Jodie (Hebrew) Admired.
Kasey (Gaelic) Vigilant.
Kassidy (Gaelic) Clever.
Kelley (Gaelic) Warrior.
Kipp (Old English) From the pointed hill.
Lindsey (Old English) From the linden tree island.
Marion (Hebrew) Bitter, graceful.
Matty (Hebrew) Gift of God.
Maxie (Latin) Most excellent.
Micah (Hebrew) Like unto the lord.
Micky (Hebrew) Like unto the Lord.
Neddy (Old English) Rich and happy protector.
Neely (Gaelic) The champion.
Niki (Greek) The people's victory.
Pace (Latin) Born at Easter or Passover.
Quin (Gaelic) Wise, intelligent.
Rowe (Gaelic) Red-haired.

Sandie (Old English) From the sandy hill.

Sloan (Gaelic) Warrior.

Terri (Greek) The harvester.

Tracee (Gaelic) Battler.

Valle (Old English) From the valley.

Vallie (Old English) From the valley.

Willie (Old German) Determined guardian.

12

Number Eight

Qualities of Number Eight

Keywords: materialistic, determined, tough, tenacious, powerful, faithful, consistent, practical, strong-willed, loyal, industrious, patient, persevering, wealthy, authoritative, eccentric, obstinate, ruthless, unscrupulous, misunderstood, cruel, greedy, morbid, destructive, vengeful.

Symbolism: The number eight represents materiality. It is feminine in nature and stands for worldly involvement. Eight is symbolized by Saturn and the double square. Its element is earth, and it is related to the balancing of spiritual and material things, cycles of time (four seasons, two solstices, and two equinoxes), the serpents of the caduceus, the mathematical symbol for infinity (eight on its side), and the joining of the two spheres of heaven and earth.

Personality: The eight vibration is much misunderstood, and like its corresponding planet Saturn, it is often feared and maligned. The number eight personality has been blessed with the ability to build on the material plane. Eights are hardworking, industrious, patient, persevering, and capable of overcoming virtually any obstacle. However, a lot of wealth and power is wielded by Eights, and that sometimes leads to ruthlessness, greed, and abuse. Eight is the number of business acumen, tenacity, and hard work, and it is associated with prosperity, power, and fame. Eights are often required to face many sorrows and losses during their lives, and their path to success may be hard and stony. In spite of their tough nature and adaptability, some Eights experience great loneliness. Their rewards come in the form of successes acquired through their own efforts and the satisfaction of accomplishing their goals. They are not the easiest people to live with, but once they have satisfied their need for security, Eights can be loyal and devoted partners.

By nature critical and jealous, Eights are always seeking proof of love and may demand constant reassurances of fidelity. They, on the other hand, have great difficulty expressing their affections and may appear cold and distant.

Careers: banker, broker, builder, cemetery worker, composer, controller, corporate lawyer, engineer, executive, financier, mathematician, miner, mortician, police officer, politician, public official, realtor, scientist.

NUMBER EIGHT NAMES—FEMALE

Adanna (Hebrew) Of the red earth.

Adonia (Greek) Beautiful.

Adorna (Latin) Adorned with jewels.

Adrianna (Latin) Black earth.

Akima (Japanese) Autumn space.

Alexandra (Greek) Helper of humankind.

Alicia (Greek) Truthful.

Alura (Old English) Divine counselor.

Amita (Israeli) Truth.

Amye (Latin) The beloved.

Anabel (Hebrew-Latin) Graceful, beautiful.

Anamarie (Hebrew) Bitter, graceful.

Annalisa (Hebrew) Graced with God's promise.

Annemarie (Hebrew) Bitter, graceful.

Aphra (Hebrew) Dust.

Aretha (Greek) Best.

Ariana (Greek) Holy one.

Artimis (Greek) Divine moon goddess.

Ashtin (Old English) From the ash tree town.

Astrid (Old Norse) Divine strength.

Athene (Greek) Wisdom.

Avril (Latin) Forthcoming.

Azucena (Arabic) Lily.

Azure (Old French) Sky blue.

Basilia (Latin) Queenly, regal.

Bea (Latin) Bringer of joy.

Beckie (Hebrew) The captivator.

Beryl (Greek) The sea-green jewel.

Beth (Hebrew) House of God.

Betsy (Hebrew) God's promise.

Bina (Hebrew) Understanding.

Brenda (Old English) Firebrand.

Brina (Old Norse) Protector.

Brunhilda (Old German) Armed battle maid.

Cadence (Greek) Pastoral simplicity and happiness.

Carla (Old German) Little woman born to command.

Carmela (Hebrew) God's vineyard.

Cassandra (Greek) Disbelieved by men.

Cayley (Greek) Pure maiden.

Celina (Greek) Moon.

Chelsea (Old English) A port of ships.

Cheryl (Old French) Little and womanly.

Clara (Latin) Bright, shining girl.

Clarabelle (French) Brilliant, beautiful.

Claribel (French) Brilliant, beautiful.

Cleo (Greek) Famous.

Colette (Greek) The people's victory.

Crystal (Latin) Without deception.

Cynthia (Greek) The moon.

Cyrilla (Greek) Mistress, ruler.

Dagmar (Old German) Glorious day.

Dahlia (Norse) From the valley.

Danielle (Hebrew) God is my judge.

Deborah (Hebrew) The bee.

Devany (Old English) Defender of Devonshire.

Docila (Latin) Gentle teacher.

Dominique (Latin) Belonging to the Lord.

Dory (Greek) Gift.

Drucilla (Latin) Strong.

Eartha (Old English) Of the earth.

Eliza (Hebrew) God's promise.

Elsbeth (Hebrew) God's promise.

Elysia (Latin) Sweetly blissful.

Erika (Old Norse) Ever-powerful, ever-ruler.

Estefania (Greek) Crowned.

Eugenia (Greek) Wellborn noble.

Faith (Old German) Belief in God.

Fawn (Old French) Young deer.

Felicidad (Latin) Joyous, fortunate.

Felicity (Latin) Joyous, fortunate.

Fleur (Latin) Flower.

Fontana (Old French) Fountain, spring.

Fontanne (Old French) Fountain, spring.

Fredericka (Old German) Peaceful ruler.

Gabrielle (Hebrew) Devoted to God.

Galina (Greek) Calm.

Georgia (Greek) Landholder, farmer.

Georgine (Greek) Landholder, farmer.

Gertrude (Old German) Spear-strength.

Gisela (Old German) Pledge or hostage.

Glenna (Old Welsh) Dweller in a valley.

Glennis (Old Welsh) Dweller in a valley.

Gloria (Latin) Glory.

Gratiana (Latin) Graceful.
Grazia (Latin) Graceful.
Greer (Latin) Watchman.
Gretchen (Latin) A pearl.
Gussie (Swedish) Staff of the Goths.
Halima (Arabic) Gentle.
Heidi (Old German) Of noble rank.
Helen (Greek) Light, a torch.
Hesper (Greek) Evening star.
Honora (Latin) Honorable.
Honoria (Latin) Honorable.
Hope (Old English) Cheerful optimism.
Ignacia (Latin) Fiery or ardent.
Ilima (Hawaiian) The flower of Oahu.
Imelda (Old German) Powerful fighter.
Inge (Norse) Hero's daughter.
Inger (Norse) Hero's daughter.
Ivory (Latin) Made of ivory.
Jacquetta (Hebrew) The supplanter.
Jana (Hebrew) The supplanter.
Janine (Hebrew) God's gift of grace.
Janis (Hebrew) God's gift of grace.
Jardena (Hebrew) Descending.
Jasmine (Arabic) Fragrant flower.
Jeanette (Hebrew) God's gift of grace.

Jenna (Arabic) A small bird.
Jocelyne (Latin) Fair and just.
Johnna (Hebrew) God's gracious gift.
Judi (Hebrew) Admired.
Julia (Greek) Young in heart and mind.
Justine (Latin) Just.
Kama (Sanskrit) Love.
Kani (Hawaiian) Sound.
Karida (Arabic) Virgin.
Karima (Arabic) Noble, exalted.
Karin (Greek) Pure maiden.
Karma (Sanskrit) Actions are fate.
Karmen (Latin) Song.
Kelula (Hebrew) Crown, laurel.
Kendra (Old English) Knowledgeable.
Keren (Greek) Pure maiden.
Kevina (Gaelic) Gentle, kind, and loving.
Kina (Hawaiian) China.
Koemi (Japanese) A little smile.
Korah (Greek) The maiden.
Kristyn (Greek) Anointed.
Kumiko (Japanese) Companion child.
Laila (Arabic) Dark as night.
Lala (Slavic) Tulip flower.
Latasha (Latin) Natal day.
Laura (Latin) Crowned with laurels.
Leah (Hebrew) Weary.

Leilani (Hawaiian) Heavenly flower.

Leonora (Greek) Light.

Lida (Greek) A woman from Lydia.

Liesel (Hebrew) God's promise.

Liliane (Latin) Lily flower.

Lilias (Latin) Lily flower.

Livia (Latin) Symbol of peace.

Lucretia (Latin) Riches, reward.

Luisa (Old German) Famous battle maid.

Madonna (Italian) My lady.

Magda (Hebrew) Woman of Magdala.

Magdalene (Hebrew) Woman of Magdala.

Marge (Latin) A pearl.

Maricela (Latin) Of the sea.

Marissa (Latin) Of the sea.

Marjorie (Latin) A pearl.

Mata (Hebrew) Gift of God.

Maude (Old German) Powerful in battle.

Mauve (French) A purple-colored mallow plant.

Meiko (Japanese) A bud.

Messina (Latin) The middle child.

Milly (Old English) Gentle strength.

Mimi (Hebrew) Bitterness.

Mireya (Latin) Wonderful.

Myrna (Gaelic) Polite, gentle.

Natalie (Latin) Natal day.

Neilla (Gaelic) The champion.

Nella (Old French) Light.

Niabi (American Indian) A fawn.

Nichola (Greek) The people's victory.

Nissa (Norse) Friendly elf or brownie.

Nita (American Indian) Bear.

Nona (Latin) The ninth.

Noreen (Latin) A rule, pattern, or precept.

Norina (Latin) A rule, pattern, or precept.

Nuri (Arabic) Light.

Nydia (Latin) From the nest.

Octavia (Latin) The eighth.

Olga (Norse) Holy.

Olwyn (Welsh) White footprint.

Opal (Sanskrit) A precious stone.

Pamelina (Greek) All honey.

Pansey (Greek) Fragrant.

Pazia (Hebrew) Golden.

Petronia (Latin) Rock or stone.

Phaedra (Greek) Bright.

Philantha (Greek) Lover of flowers.

Phylis (Greek) A green branch.

Pia (Latin) Devout.

Pierette (Latin) Rock or stone.

Polly (Latin) Little.

Pris (Latin) From ancient times.

Quintessa (Latin) Fifth essence.

Rafaela (Hebrew) Healed by God.

Ramona (Old German) Mighty or wise protector.

Raphaela (Hebrew) Healed by God.

Raylina (Hebrew) Innocent as a lamb.

Reanna (Welsh) A witch.

Reba (Hebrew) The captivator.

Renelle (Latin) Reborn.

Reta (African) To shake.

Reynalda (Old English) Mighty and powerful.

Rianon (Welsh) A witch.

Rosa (Greek) A rose.

Rosabelle (French-Greek) Beautiful rose.

Rosaleen (Spanish) Beautiful rose.

Rosetta (Greek) A rose.

Rue (English) Regret.

Samara (Hebrew) Ruled by God.

Samaria (Hebrew) Ruled by God.

Sarina (Hebrew) Of royal status.

Savannah (Spanish) Barren.

Scarlett (Old English) Scarlet-colored.

Serena (Latin) Calm, serene.

Sheba (Hebrew) From Sheba.

Sienna (Latin) A brownish-red color.

Simona (Hebrew) One who hears.

Simonne (Hebrew) One who hears.

Solana (Spanish) Sunshine.

Sondra (Greek) Helper of humankind.

Starla (English) Star.

Sula (Icelandic) Large sea bird.

Summer (English) Summer season.

Susanna (Hebrew) Graceful lily.

Suzie (Hebrew) Graceful lily.

Sybilla (Greek) Prophetess.

Syndi (Greek) The moon.

Tabatha (Greek) Gazelle.

Taima (American Indian) Crash of thunder.

Talitha (Hebrew) Child.

Tama (Japanese) Well-polished.

Tamra (Hebrew) Palm tree.

Tani (Greek) Giant.

Tavia (Latin) The eighth.

Tempest (Old French) Stormy.

Tera (Japanese) Calm.

Terra (Latin) The planet Earth, land.

Therese (Greek) The harvester.

Thora (Old Norse) Thunder.

Tina (Greek) Anointed.

Toshi (Japanese) Mirror reflection.

Trina (Greek) Pure maiden.

Ulane (Polynesian) Cheerful.

Unity (Old English) Unity.

Valentina (Latin) Strong.
Velda (Sanskrit) Wise.
Velma (Old German)
   Determined guardian.
Virginia (Latin) Virginal.
Wendy (Old German) The
   wanderer.
Winter (Old English) Winter.
Xandra (Greek) Helper of
   humankind.

Xaviera (Arabic) Bright.
Xena (Greek) Hospitable.
Xenia (Greek) Hospitable.
Xylia (Greek) Of the wood.
Yetta (Old English) Giver.
Zabrina (Latin) A princess.
Zelia (Greek) Zealous.
Zerlinda (Hebrew) Beautiful as
   the dawn.

nAMES nUMBER EIGHT—MALE

Abdulla (Arabic) Servant of
   God.
Abe (Hebrew) My father is joy.
Abraham (Hebrew) Father of
   the multitude.
Abram (Hebrew) The lofty one
   is father.
Adolfo (Old German) Noble
   wolf.
Adolpho (Old German) Noble
   wolf.
Adonis (Greek) Handsome.
Alaric (Old German) Rules all.
Albion (Celtic) White cliffs.
Alcott (Old English) From the
   cottage.

Alejandro (Greek) Helper of
   humankind.
Alistair (Greek) Helper of
   humankind.
Allen (Gaelic) Handsome,
   cheerful.
Alton (English) From the old
   town.
Alvah (Latin) White, fair.
Amery (Old German)
   Industrious ruler.
Angus (Celtic) Unique strength.
Antony (Latin) Praiseworthy,
   without a peer.
Apollo (Greek) Manly.
Archer (Old English) Bowman.

Archie (Old German) Genuinely bold.

Armstrong (Old English) Strong arm.

Artie (Celtic-Welsh) Noble, bear-hero.

Ashford (Old English) From the ash tree ford.

August (Latin) Majestic dignity.

Bainbridge (Old English) Bridge over white water.

Barclay (Old English) From the birch tree meadow.

Barlow (Old English) From the bare hill.

Barnett (Old English) Nobleman.

Baruch (Hebrew) Blessed.

Benedict (Latin) Blessed.

Bennett (Latin) Blessed.

Bernard (Old German) Brave as a bear.

Berne (Old German) Brave as a bear.

Bernie (Latin) Brave as a bear.

Bevan (Celtic) Youthful warrior.

Bill (Old German) Determined guardian.

Boaz (Hebrew) In the Lord is strength.

Bond (Old English) Tiller of the soil.

Braden (Old English) From the broad valley.

Brexton (Gaelic) Freckled.

Brian (Celtic) Strength with virtue and honor.

Brickman (Old English) Bridge man.

Brodie (Gaelic) A ditch.

Brodrick (Old English) From the bridge road.

Bryant (Celtic) Strength with virtue and honor.

Bryce (Celtic) Quick.

Burl (Old English) Cupbearer.

Callum (Gaelic) Mild, gentle.

Camilo (Latin) Young ceremonial attendant.

Carsten (Greek) Anointed.

Casper (Persian) Treasurer.

Chadwick (Old English) From the warrior's town.

Chauncey (Old English) Chancellor.

Cicero (Latin) Chickpea.

Clemens (Latin) Merciful.

Cletus (Greek) Famous.

Clovis (Latin) Renowned fighter.

Cole (Greek) The people's victory.

Colin (Gaelic) Strong and virile.

Constantine (Latin) Firmness, constancy.

Cornelius (Latin) Yellow, horn-colored.

Creed (Latin) Belief, guiding principle.

Crispian (Latin) Curly-haired.

Curt (Old French) Courteous.

Dafydd (Hebrew) Beloved.

Dante (Latin) Lasting.

Delmar (Old French) From the sea.

Destin (Old French) Destiny.

Dewey (Hebrew) Beloved.

Dudley (Old English) From the people's meadow.

Duffy (Celtic) Dark.

Durand (Latin) Enduring.

Durward (Old English) Gatekeeper, door warden.

Dwight (Old English) A cutting or clearing.

Eamonn (Old English) Prosperous guardian.

Eben (Hebrew) Stone.

Ebenezer (Hebrew) Stone of help.

Edgar (Old English) Prosperous spearman.

Efrain (Hebrew) Very fruitful.

Elbert (Old German) Bright, famous.

Eli (Hebrew) Jehovah.

Elmer (Old English) Noble, famous.

Elvin (Old English) Elf-friend.

Emanuel (Hebrew) God is with us.

Emerson (Old German) Son of the industrious ruler.

Emrys (Greek) Immortal.

Enos (Hebrew) Man.

Enrique (Old German) Lord of the manor.

Eric (Old Norse) Ever-powerful, ever-ruler.

Erroll (Old English) Nobleman.

Fabrice (Latin) Craftsman.

Faine (Old English) Good-natured.

Fane (Old English) Good-natured.

Farand (Old German) Pleasant and attractive.

Farris (Old English) Iron-strong.

Favian (Latin) Bean grower.

Favio (Latin) Bean grower.

Felipe (Greek) Lover of horses.

Findlay (Gaelic) Little fair-haired soldier.

Finley (Gaelic) Little fair-haired soldier.

Floyd (Old Welsh) Grey-haired.

Flynn (Gaelic) Son of the red-haired man.

Freeman (Old English) Free.

Gallagher (Gaelic) Eager helper.

Garfield (Old English) Triangular field.

Garrett (Old German) Spear-mighty.

Garvin (Old German) Spear-friend.

Gaspar (Persian) Treasurer.

Gavin (Old Welsh) White hawk.

Gaynor (Gaelic) Son of the fair-haired man.

Gentry (Old French) Wellborn.

Gerard (Old English) Spear-strong.

Gerrard (Old English) Spear-strong.

Giuseppe (Hebrew) He shall add.

Glendon (Scottish Gaelic) From the valley fortress.

Goddard (Old German) Divinely firm.

Godfrey (Old German) God's divine peace.

Gothart (Old German) Divinely firm.

Grahame (Old English) The grey home.

Gregg (Latin) Watchman.

Griswold (Old German) From the grey forest.

Guy (Old German) Warrior.

Hale (Old English) Hero.

Hamid (Arabic) The praised one.

Hartley (Old English) From the deer meadow.

Hassan (Arabic) Good-looking.

Hayward (Old English) From the hedged forest.

Horst (Old German) A thicket.

Horten (Old English) from the grey estate.

Hugh (Old English) Intelligence.

Hurley (Gaelic) Sea tide.

Hutton (Old English) From the house on the ledge.

Ingram (Old Norse) The son's raven.

Isadore (Greek) Gift of Isis.

Jadon (Hebrew) Jehovah has heard.

Jefferson (Old English) Son of Jeffery.

Jerardo (Old English) Spear-strong.

Jeremias (Hebrew) Exalted by the Lord.

Josiah (Hebrew) Jehovah supports.

Jovan (Latin) Father of the sky.

Judah (Hebrew) Praised.

Julien (Latin) Related to Julius.

Justis (Latin) Just.

Kaelan (Gaelic) Powerful in battle.

Kareem (Arabic) Noble, exalted.

Kasimir (Slavic) Enforces peace.

Keiji (Japanese) Director.

Keith (Celtic) From the forest.

Kenn (Gaelic) Handsome.

Kenrick (Old English) Bold ruler.

Kerwin (Gaelic) Little, jet-black one.

Khalil (Arabic) Friend.

Kian (Hebrew) God's gracious gift.

Knute (Old Norse) Knot.

Koby (Hebrew) The supplanter.

Koi (Hawaiian) Urge, implore.

Krishna (Hindi) Delightful.

Kurtis (Old French) Courteous.

Kyle (Gaelic) Handsome.

Laird (Scottish) Landed proprietor.

Lambert (Old German) Bright land.

Lance (Old French) Servant, attendant.

Lazarus (Hebrew) God will help.

Liam (Old German) Determined guardian.

Lockwood (Old English) From the enclosed forest.

Louie (Old German) Famous warrior.

Mac (Scottish Gaelic) Son of.

Malvin (Gaelic) Brilliant chief.

Marc (Latin) Follower of Mars.

Mariano (Hebrew) Bitter, graceful.

Mason (Old French) Stoneworker.

Mathias (Hebrew) Gift of God.

Mauricio (Latin) Dark-skinned.

Mayer (Hebrew) Bringer of light.

Mendel (Yiddish) Comforter.

Mercer (Old English) Merchant, storekeeper.

Merlin (Old English) Falcon or hawk.

Mischa (Hebrew) Like unto the Lord.

Mitch (Hebrew) Like unto the Lord.

Moises (Hebrew) Drawn out of the water.

Monroe (Gaelic) From the mouth of the Roe River.

Monti (Latin) From the pointed mountain.

Mortie (Old French) Still water.

Moses (Hebrew) Drawn out of the water.

Nat (Hebrew) Gift of God.

Natale (Latin) Natal day.

Navarro (Old Spanish) Plains.

Neall (Gaelic) The champion.

Nye (Welsh) Honor or gold.

Obed (Hebrew) Servant of God.

Obediah (Hebrew) Servant of God.

Ollie (Latin) Symbol of peace.

Orion (Latin) Dawning, golden.

Osman (Old English) God's servant.

Osmond (Old English) Divine protector.

Owain (Old Welsh) Wellborn young warrior.

Paco (Latin) Free.

Parrish (Old English) From the churchyard.

Perrie (Old French) Pear tree.

Pierre (Latin) Rock or stone.
Pincas (Hebrew) Oracle.
Prescott (Old English) From the priest's cottage.
Preston (Old English) From the priest's estate.
Primo (Latin) First, first child.
Quincy (Old French) From the fifth son's estate.
Raimondo (Old German) Mighty or wise protector.
Rajan (Sanskrit) King.
Randall (Old English) Shield-wolf.
Ransom (Old English) Son of the shield.
Renjiro (Japanese) Clean, upright, honest.
Reynato (Old English) Mighty and powerful.
Rhett (Welsh) Enthusiastic.
Riccardo (Old German) Wealthy, powerful.
Roane (Gaelic) Red-haired, red.
Rob (Old English) Bright fame.
Ronan (Gaelic) Little seal.
Ross (Celtic) From the peninsula.
Royal (Old French) Royal.
Royd (Old Norse) From the forest clearing.
Rupert (Old English) Bright fame.
Rutger (Old German) Famous spearman.

Salomon (Hebrew) Peaceful.
Samuel (Hebrew) His name is God.
Sanders (Old English) Son of Alexander.
Saul (Hebrew) Asked for.
Selwyn (Old English) Friend from the manor house.
Sergius (Latin) The attendant.
Seymour (Old French) From the town of Saint Maur.
Shepard (Old English) Shepherd.
Sherwood (Old English) From the bright forest.
Sig (Old German) Victorious protector.
Spence (Old English) Dispenser of provisions.
Spencer (Old English) Dispenser of provisions.
Stafford (Old English) From the riverbank landing.
Stefano (Greek) Crowned.
Steve (Greek) Crowned.
Stoney (Old English) Stone.
Sully (Gaelic) Black-eyed.
Tadashi (Japanese) Serves the master faithfully.
Tavis (Hebrew) The devoted brother.
Teiji (Japanese) Righteous second son.
Terencio (Latin) Smooth.
Theon (Greek) Godly man.

Theron (Greek) The hunter.

Thorne (Old English) From the thorny embankment.

Timmy (Greek) Honoring God.

Timon (Greek) Honoring God.

Titus (Greek) Of the giants.

Travis (Old French) At the crossroads.

Trevor (Gaelic) Wise and discreet.

Tyler (Old English) Maker of tiles.

Ulrich (Old German) Wolf-ruler.

Urbano (Latin) From the city.

Vladamir (Old Slavic) Powerful prince.

Wat (Old German) Mighty warrior.

Wayland (Old English) From the land by the road.

Waylen (Old English) From the land by the road.

Wells (Old English) From the springs.

Wendall (Old German) The wanderer.

Wilbert (German-French) Brilliant hostage.

Wilmer (Old English) Determined guardian.

Wyatt (Old French) Little warrior.

Xanthus (Latin) Golden-haired.

Ximenes (Hebrew) One who hears.

Xiomar (Old German) Famous in battle.

Yves (Norse) Battle archer.

Zeus (Greek) Living.

NUMBER EIGHT NAMES—UNISEX

Abbey (Hebrew) My father is joy.

Andy (Greek) Strong and manly.

Austen (Latin) Majestic dignity.

Avery (Old English) Good counselor.

Barrie (Celtic) One whose intellect is sharp.

Beverly (Old English) From the beaver meadow.

Bobbie (Old English) Bright fame.

Brooks (Old English) From the brook.

Casey (Gaelic) Vigilant.

Cassidy (Gaelic) Clever.

Daryn (Gaelic) Great.

Denny (Greek) God of wine.

Dusty (Old German) Valiant fighter.

Freddy (Old German) Peaceful ruler.

Gaby (Hebrew) Devoted to God.

Gerrie (Old German) Spear-mighty.

Jess (Hebrew) Wealthy.

Jordan (Hebrew) Down-flowing.

Kiley (Gaelic) Handsome.

Lauren (Latin) Crowned with laurels.

Leslie (Scottish Gaelic) From the grey fortress.

Lonny (Old English) Solitary.

Lyall (Old French) From the island.

Merle (French) Blackbird.

Nicky (Greek) The people's victory.

Nicol (Greek) The people's victory.

Nika (Greek) The people's victory.

Padget (French) Useful assistant.

Randy (Old English) Swift wolf.

Ray (Old French) Kingly.

Robby (Old English) Bright fame.

Rody (Latin) Dew of the sea.

Rowan (Gaelic) Red-haired.

Sammy (Hebrew) His name is God.

Shandy (Old English) Rambunctious.

Shay (Gaelic) From the fairy fort.

Shelby (Old English) From the estate on the ledge.

Stevie (Greek) Crowned.

Temple (Old English) From the town of the temple.

Toby (Hebrew) The Lord is good.

Vail (Old English) From the valley.

Val (Old English) From the valley.

Wesley (Old English) From the western meadow.

13

number nine

Qualities of number nine

Keywords: humanitarian, idealistic, broad-minded, generous, energetic, bold, courageous, competitive, inspirational, independent, strong, charitable, artistic, resourceful, spiritual, successful, strong-willed, global-minded, well-traveled, impulsive, hasty, fanatical, ill-tempered, argumentative, conceited, self-centered, deceptive.

Symbolism: The number nine represents accumulation. It is the greatest of all numbers because it contains the qualities of all the others. It is masculine in nature and stands for universal love. Nine is symbolized by Mars and the scepter and the orb. Its element is fire, and it is related to the muses, consciousness, mental and spiritual attainment, and the nine months of pregnancy.

Personality: The number nine individual is a true humanitarian, capable of great spiritual and mental achievement. Nines can become leaders in any fields they choose. They have the ability to understand the human condition, and in their work with others they tend to be both courageous and selfless. Nines are born fighters and have no fear of anything or anyone. They will not tolerate injustice and are often in the forefront of the battle against discrimination and inequity. These servers of mankind may demonstrate their love for other human beings by becoming physicians, evangelists, missionaries, or soap box orators. Nines are both inspired and inspiring. However, they tend to be impulsive, and many are accident prone. Since they pursue their ambitions with much grit and determination, they often make enemies along the way. Some Nines are exceptionally gifted artistically. Whatever the profession, Nines usually attempt to sway others to contribute to the goodness of life. Their devotion to their loved ones is so great that they will sacrifice everything for them.

Nines can be very impatient and often forget to think before they speak. Sometimes they become so sentimental that the potential for good is dissipated and lost in aimless dreaming. When they allow their hearts to totally rule their heads, Nines can become overly sympathetic or fanatical and may end up giving away everything they own.

Careers: athlete, barber, ballistics expert, doctor, entrepreneur, evangelist, fireman, ironworker, journalist, judge, metalworker, military officer, musician, minister, occultist, politician, priest, publisher, social worker, spiritual leader, steelworker, surgeon, union leader, travel expert or guide, TV or radio personality, writer.

NUMBER NINE NAMES—FEMALE

Acacia (Greek) Honorable.

Adele (Old German) Of noble rank.

Alda (Latin) Elder.

Alida (Greek) Beautifully dressed.

Alima (Arabic) Learned in music and dance.

Allena (Greek) Light, a torch.

Alma (Latin) Soul, spirit.

Almeda (Latin) Pressing toward the goal.

Almira (Arabic) Exalted.

Alva (Latin) White, fair.

Alvern (Latin) Spring, greening.

Angelina (Greek) Heavenly messenger.

Angharad (Welsh) Free from shame.

Anissa (Hebrew) Full of grace.

Anita (Hebrew) Full of grace.

Anjanette (Hebrew) God's gift of grace.

Annamaria (Hebrew) Bitter, graceful.

Annora (Old English) Light and graceful.

Anona (Latin) Yearly crops.

Areta (Greek) Best.

Augusta (Latin) Majestic dignity.

Autumn (Latin) The fall season.

Aya (Japanese) Patterned cloth, damask.

Aziza (Arabic) Beloved.

Bambi (Italian) Child.

Beatrice (Latin) Bringer of joy.

Beatriz (Latin) Bringer of joy.

Bekky (Hebrew) The captivator.

Belle (French) Beautiful.

Benedicta (Latin) Blessed.

Bernadine (Old German) Brave as a bear.

Bertha (Old German) Shining.

Bess (Hebrew) God's promise.

Betty (Hebrew) God's promise.

Blanche (Old French) White, fair.

Blondelle (French) Little fair one.

Blythe (Old English) Joyful.

Breezy (English) Lightly windy.

Brenna (Gaelic) Little raven.

Briana (Celtic) Strength with virtue and honor.

Brianne (Celtic) Strength with virtue and honor.

Brigitte (Celtic) Strong and mighty.

Brydie (Celtic) Strong and mighty.

Bryony (Celtic) Strength with virtue and honor.

Calandra (Greek) Lark.

Candida (Greek) Glittering, glowing white.

Carlotta (Old German) Little woman born to command.

Carmen (Latin) Song.

Carole (Old German) Noble and strong.

Carrie (Old German) Little woman born to command.

Catriona (Greek) Pure maiden.

Ceiridwen (Welsh) Goddess of bards.

Chantel (Latin) Song.

Charmaine (Latin) Song.

Charo (Spanish) Rosary.

Coleen (Gaelic) Girl.

Concetta (Latin) Conception.

Constancia (Latin) Firmness, constancy.

Consuela (Spanish) Consolation.

Cybill (Greek) A prophetess.

Dacia (Greek) From Dacia.

Debbie (Hebrew) The bee.

Deidre (Gaelic) Sorrow.

Deirdre (Gaelic) Sorrow.

Destinee (Old French) Destiny.

Dextra (Latin) Skillful, adept.

Dinah (Hebrew) Judged.

Dominga (Latin) The Lord's day.

Dulce (Latin) Sweet and charming.

Dulcie (Latin) Sweet and charming.

Elberta (Old English) Noble and brilliant.

Eleni (Greek) Light, a torch.

Elisabeth (Hebrew) God's promise.

Elisha (Hebrew) God is salvation.

Elly (Greek) Light.

Elvina (Old English) Elf-friend.

Emmaline (Old German) One who leads the universe.

Erica (Old Norse) Ever-powerful, ever-ruler.

Evangelina (Greek) Bearer of good tidings.

Eveline (Hebrew) Life-giving.

Fawna (Old French) Young deer.

Felicia (Latin) Joyous, fortunate.

Fernanda (Old German) World-daring.

Fidelity (Latin) Faithful.

Fiona (Gaelic) Fair.

Florimel (French-Greek) Dark flower.

Francoise (Latin) Free.

Gaia (Greek) The earth.

Gena (French) Pure white wave.

Geneva (French) Pure white wave.

Genevra (French) Pure white wave.

Gloriane (Latin) Glory.

Goldy (Old English) Golden.

Gwenda (Old Welsh) White-browed.

Gwendoline (Old Welsh) White-browed.

Harlene (Old English) From the long field.

Hattie (French) Mistress of the household.

Helena (Greek) Light, a torch.

Hinda (Hebrew) Female deer.

Horatia (Latin) Keeper of the hours.

Hosanna (Greek) A prayer of acclamation and praise.

Idalia (Greek) Behold the sun.

Ilene (Greek) Light, a torch.

Ilse (Hebrew) God's promise.

Imogen (Latin) Image.

Inez (Greek) Pure, chaste.

Ivette (Old French) Archer with a yew bow.

Jacquelyn (Hebrew) The supplanter.

Jeannine (Hebrew) God's gift of grace.

Jennifer (French) Pure white wave.

Jesenia (Arabic) Flower.

Jinny (Latin) Virginal.

Jiselle (Old German) Pledge or hostage.

Joann (Hebrew) God's gift of grace.

Johanna (Hebrew) God's gift of grace.

Jonina (Israeli) Little dove.

Jordana (Hebrew) Downflowing.

Judith (Hebrew) Admired.

Kalil (Arabic) Beloved.

Kaliliah (Arabic) Beloved.

Kamaria (African) Moonlike.

Karina (Greek) Pure maiden.

Keeley (Gaelic) Beautiful.

Keiki (Hawaiian) Child.

Keitha (Celtic) From the forest.

Kesia (African) Favorite.

Klarissa (Latin) Bright, shining girl.

Kora (Greek) The maiden.

Lani (Hawaiian) Sky.

Lea (Hebrew) Weary.

Lian (Chinese) The graceful willow.

Lilianna (Latin) Lily flower.

Lilli (Latin) Lily flower.

Lina (Greek) Light, a torch.

Linnette (Old French) Songbird.

Lise (Hebrew) God's promise.

Lisette (Hebrew) God's promise.

Lori (Latin) Crowned with laurels.

Lottie (French) Little and womanly.

Louise (Old German) Famous battle maid.

Lucrecia (Latin) Riches, reward.

Luella (Old English) Elf.

Luellen (Old English) Elf.

Madeline (Hebrew) Woman of Magdala.

Maeko (Japanese) Truth child.

Magnolia (French) Flower named after Pierre Magnol.

Mahala (Hebrew) Affection.

Mahalia (Hebrew) Affection.

Maisey (Latin) A pearl.

Malia (Hebrew) Bitterness.

Malvina (Greek) Soft and tender.

Marcia (Latin) Follower of Mars.

Margo (Latin) A pearl.

Marguerite (Latin) A pearl.

Marla (Hebrew) Bitterness.

Maura (Hebrew) Bitterness.

Meagen (Gaelic) Soft, gentle.

Melina (Latin) Canary-yellow-colored.

Melodie (Greek) Song.

Mercedes (Spanish) Compassion.

Miriam (Hebrew) Bitterness.

Mohala (Hawaiian) Petals unfolding, shining forth.

Morag (Celtic) Great.

Moselle (Hebrew) Drawn out of the water.

Moyra (Celtic) Great.

Musetta (French) Child of the muses.

Mystica (French) Air of mystery.

Nadya (Russian) Hope.

Neile (Gaelic) The champion.

Nelda (Old English) Born under an elder tree.

Nicola (Greek) The people's victory.

Noelle (French) Born at Christmas.

Odessa (Greek) Wandering, quest.

Olive (Latin) Symbol of peace.

Oona (Latin) Unity.

Opaline (Sanskrit) A precious stone.

Orianna (Latin) Dawning, golden.

Ottilie (Old German) Prosperous.

Paola (Latin) Little.

Pennie (Greek) The weaver.

Perina (Latin) Rock or stone.

Phebe (Greek) Shining.

Phillippa (Greek) Lover of horses.

Phyllys (Greek) A green branch.

Poala (Latin) Little.

Priscilla (Latin) From ancient times.

Prospera (Latin) Favorable.

Prunella (Latin) Plum color.

Queena (Old English) Queen.

Ranita (Hebrew) Song.

Raya (Israeli) Friend.

Regina (Latin) Queen.

Renate (Latin) Reborn.

Rexana (Latin) Regal and graceful.

Rhetta (African) To shake.

Roanna (Greek-Hebrew) A rose, full of grace.

Robinette (Old English) Bright fame.

Rosabel (French-Greek) Beautiful rose.

Rosalynd (Spanish) Beautiful rose.

Rosamond (Old German) Famous guardian.

Rosemaria (English) Rose of Saint Mary.

Ruthie (Hebrew) Compassionate.

Samuela (Hebrew) His name is God.

Satin (French) Satin fabric.

Shelia (Celtic) Musical.

Shirleen (Old English) From the bright meadow.

Shirlene (Old English) From the bright meadow.

Sibley (Greek) Prophetess.

Silvia (Latin) From the forest.

Sophie (Greek) Wisdom.

Sorrell (Old French) Reddish-brown.

Stormie (English) Stormy.

Sue (Hebrew) Graceful lily.

Sukey (Hebrew) Graceful lily.

Tacita (Latin) To be silent.

Tamara (Hebrew) Palm tree.

Tammy (Hebrew) Perfection.

Tania (Greek) Giant.

Teena (Greek) Anointed.

Tess (Greek) The harvester.

Thalassa (Greek) From the sea.

Thyra (Greek) Shield-bearer.

Tiana (Russian) Fairy queen.

Tiffany (Greek) Appearance of God.

Trenna (Greek) Pure maiden.

Twyla (Old English) Woven of a double thread.

Ulrika (Old German) Wolf-ruler.

Una (Latin) Unity.

Ursala (Latin) Little bear.

Vala (Old German) The chosen one.

Valerie (Latin) Strong.

Vanessa (Greek) Butterflies.

Venus (Latin) Venus.

Verity (Latin) Truth.

Veronique (Greek) Bringer of victory.

Vicki (Latin) Victorious.

Vida (Sanskrit) Wise.

Wenona (American Indian) Firstborn daughter.

Wynne (Celtic) Fair white maiden.

Xanthe (Greek) Golden yellow.

Xiomara (Old German) Famous in battle.

Yasmin (Arabic) Jasmine flower.

Ynes (Greek) Pure, chaste.

Yolanda (Greek) Violet flower.

Zenobia (Arabic) Her father's ornament.

Zetta (Old English) Sixth-born.

Zeva (Greek) Sword.

Zia (Latin) Kind of grain.

Zola (Italian) Ball of earth.

NUMBER NINE NAMES—MALE

Absalom (Hebrew) Father of peace.

Ace (Latin) Unity.

Adlai (Hebrew) My witness.

Alastair (Greek) Helper of humankind.

Alden (Old English) Old wise protector.

Aldis (Old German) Old and wise.

Aldous (Old German) Old and wise.

Alessandro (Greek) Helper of humankind.

Alick (Greek) Helper of humankind.

Allon (Gaelic) Handsome and cheerful.

Alphonse (Old German) Noble and eager.

Alvis (Old English) All-knowing.

Ambrosius (Greek) Immortal.

Amory (Old German) Industrious ruler.

Anderson (Greek) Strong and manly.

Angelo (Greek) Heavenly messenger.

Ansell (Old French) Adherent of a nobleman.

Anson (Old German) Of divine origin.

Arney (Old German) Strong as an eagle.

Arsenio (Greek) Potent.

Auden (Old English) Old friend.

Audwin (Old German) Noble friend.

Augustine (Latin) Majestic dignity.

Aurelio (Latin) Golden.

Baran (Russian) Forceful, virile.

Barnaby (Greek) Son of prophecy.

Bert (Old English) Bright.

Bogart (Old German) Strong as a bow.

Boris (Slavic) Battler, warrior.

Bowie (Gaelic) Yellow-haired.

Brando (Old English) Firebrand.

Bret (Celtic) Native of Brittany.

Bud (Old English) Herald, messenger.

Buddie (Old English) Herald, messenger.

Burdett (Old French) Little shield.

Burton (Old English) From the fortress.

Butch (Old English) Bright.

Caden (Celtic) Battle spirit.

Caelan (Gaelic) Powerful in battle.

Cain (Hebrew) Possession or possessed.

Caldwell (Old English) Cold spring.

Calvert (Old English) Herdsman.

Caradoc (Celtic) Beloved.

Carmine (Latin) Song.

Carrick (Scottish) From the rugged hills.

Casimir (Slavic) Proclamation of peace.

Charley (Old German) Strong and masculine.

Chase (Old French) Hunter.

Chet (Old English) From the fortified camp.

Chevy (Old French) Horseman, knight.

Chilton (Old English) From the farm by the spring.

Clark (Old French) Scholar.

Clayton (Old English) From the town built on clay.

Cleavon (Old English) From the cliff.

Clement (Latin) Merciful.

Cliff (Old English) From the cliff.

Coleman (Old English) Nicholas's man.

Conroy (Gaelic) Wise one.

Conway (Gaelic) Hound of the plain.

Cooper (Old English) Barrelmaker.

Cormick (Gaelic) Charioteer.

Courtland (Old English) From the court land.

Creighton (Old English) From the town near the creek.

Culbert (Old English) Brilliant seafarer.

Culver (Old English) Dove.

Curtis (Old French) Courteous.

Cynric (Old English) From the royal line of kings.

Darius (Greek) Wealthy.

Dayan (Hebrew) Judge.

Delmon (Old English) Of the mountain.

Delmore (Old French) From the sea.

Denton (Old English) Valley town.

Dewitt (Flemish) White or blond.

Dick (Old German) Wealthy, powerful.

Duane (Gaelic) Dark.

Dugal (Celtic) From the dark stream.

Dwayne (Gaelic) Dark.

Earl (Old English) Nobleman.

Eddie (Old English) Prosperous guardian.

Edsel (Old English) Rich man's house.

Egan (Gaelic) Ardent, fiery.

Ekon (Nigerian) Strong.

Elijah (Hebrew) Jehovah is God.

Elmo (Italian) Helmet.

Elwin (Old English) Elf-friend.

Emilio (Latin) Flattering.

Enoch (Hebrew) Experienced, dedicated.

Erhard (Old German) Strong honor.

Ernest (Old English) Earnest.

Erskine (Scottish Gaelic) From the height of the cliff.

Etienne (Greek) Crowned.

Ewald (Old English) Law, powerful.

Fabricio (Latin) Craftsman.

Faraji (Swahili) Consolation.

Farquhar (Gaelic) Friendly man.

Farrell (Gaelic) Most valorous.

Felton (Old English) From the estate on the meadow.

Fidel (Latin) Faithful.

Filbert (Old English) Very brilliant.

Filmer (Old English) Very famous.

Fitzgerald (Old English) Son of the spear-mighty.

Fredric (Old German) Peaceful ruler.

Garner (Old French) Armed sentry.

Garth (Old Norse) Groundskeeper.

Gavan (Old German) Spear-friend.

Geordie (Greek) Landholder, farmer.

Gervais (Old German) Honorable.

Giacomo (Hebrew) The supplanter.

Gianni (Hebrew) God's gracious gift.

Gideon (Hebrew) Feller of trees.

Gino (Hebrew) God's gracious gift.

Giomar (Old German) Famous in battle.

Godwin (Old English) Friend of God.

Gonzalo (Old German) Complete warrior.

Gorden (Old English) Round hill.

Grayson (Old English) Son of the bailiff.

Hari (Hindi) Princely.

Harlan (Old English) From the army land.

Henri (Old German) Lord of the manor.

Hewett (Old French-German) Little and intelligent.

Hilario (Latin) Cheerful.

Hogan (Gaelic) Youthful.

Holmes (Old English) From the river islands.

Horton (Old English) From the grey estate.

Hudson (Old English) Son of the hoodsman.

Hunt (Old English) A hunter.

Irvin (Old English) Sea friend.

Jabin (Hebrew) God has built.

Jacinto (Greek) Hyacinth flower.

Jake (Hebrew) God's gracious gift.

Jakeem (Arabic) Raised up.

Jamil (Arabic) Handsome.

Jareb (Hebrew) He will contend.

Jay (Old French) Blue jay.

Jeff (Old German) God's divine peace.

Jeronimo (Latin) Sacred or holy name.

Job (Hebrew) The afflicted.

Justino (Latin) Just.

Kaori (Japanese) Add a man's strength.

Kassim (Arabic) Divided.

Keane (Old English) Quick, brave.

Keneth (Gaelic) Handsome.

Kenley (Old English) From the royal meadow.

Keon (Hebrew) God's gracious gift.

Keoni (Hebrew) God's gracious gift.

Kiefer (Gaelic) Cherished, handsome.

Kiernan (Gaelic) Little and dark-skinned.

Konrad (Old German) Bold counselor.

Kwasi (African) Born on Sunday.

Lamar (Old German) Famous throughout the land.

Lander (Old English) Owner of the grassland.

Lathrop (Old English) From the barn farmstead.

Laurens (Latin) Crowned with laurels.

Lawrence (Latin) Crowned with laurels.

Leighton (Old English) From the meadow farm.

Leith (Celtic) Broad wide river.

Les (Latin) From the chosen camp.

Lorimer (Old English) Saddle-maker.

Lyle (Old French) From the island.

Madoc (Welsh) Fortunate.

Malachy (Greek) Strong and manly.

Marius (Latin) Follower of Mars.

Marshal (Old French) Steward, horse-keeper.

Marsten (Old English) From the town near the marsh.

Martell (Old French) Mace.

Mateo (Hebrew) Gift of God.

Matt (Hebrew) Gift of God.

Matthew (Hebrew) Gift of God.

Maxwell (Old English) Large well or spring.

Mayo (Gaelic) From the plain of the yew trees.

Mayor (Latin) Greater.

Meir (Hebrew) Bringer of light.

Melville (Old French) From the estate of the hard worker.

Mervin (Old English) Lover of the sea.

Mick (Hebrew) Like unto the Lord.

Minoru (Japanese) Bear fruit.

Mohammed (Arabic) Praised.

Munro (Gaelic) From the mouth of the Roe River.

Nehemiah (Hebrew) Compassion of Jehovah.

Nicholas (Greek) The people's victory.

Norbie (Old Norse) Brilliant hero.

Norry (Old French) Northerner.

Ogden (Old English) From the oak valley.

Oliver (Latin) Symbol of peace.

Olivier (Latin) Symbol of peace.

Orsino (Latin) Bearlike.

Orson (Latin) Bearlike.

Otis (Greek) Keen of hearing.

Ozzie (Old English) Divine spearman.

Packston (Latin) From the peaceful town.

Paine (Latin) Country man.

Patrice (Latin) Wellborn, noble.

Paxton (Latin) From the peaceful town.

Penrod (Old German) Famous commander.

Phil (Greek) Lover of horses.

Phineas (Hebrew) Oracle.

Prentice (Old English) Apprentice.

Quint (Latin) Fifth child.

Rafferty (Gaelic) Rich and prosperous.

Ramsey (Old English) From the ram's island.

Ranger (Old French) Guardian of the forest.

Ransell (African) Borrowed all.

Ranulf (Old English) Shield-wolf.

Ravid (Hindi) Sun.

Raviv (Hindi) Sun.

Rayburn (Old English) From the deer brook.

Raymond (Old German) Mighty or wise protector.

Red (Old English) Red-colored.

Reece (Welsh) Enthusiastic.

Refugio (Spanish) Refuge, shelter.

Reid (Old English) Red-haired.

Rexford (Old English) King's river-crossing.

Reyes (Latin) Kings.

Richy (Old German) Wealthy, powerful.

Rico (Old German) Wealthy, powerful.

Rider (Old English) Horseman.

Ringo (Old English) Ring.

Ritchie (Old German) Wealthy, powerful.

Robertson (English) Son of Robert.

Rockwell (Old English) From the rocky spring.

Roderic (Old German) Famous ruler.

Rodney (Old English) From the island clearing.

Rogelio (Old German) Famous spearman.

Roger (Old German) Famous spearman.

Roka (Japanese) White crest of the wave.

Rollo (Old German) From the famous land.

Ruddy (Old English) From the red enclosure.

Rutland (Old Norse) From the stumpland.

Sabin (Latin) Sabine man.

Salim (Arabic) Safe, peace.

Salvator (Italian) Savior.

Samson (Hebrew) Like the sun.

Sandford (Old English) From the sandy hill.

Sauncho (Latin) Sanctified.

Sebastian (Latin) Venerated.

Selby (Old English) From the village by the mansion.

Serge (Latin) The attendant.

Sergei (Latin) The attendant.

Sextus (Latin) Sixth-born.

Shamus (Hebrew) The supplanter.

Shaun (Hebrew) God's gracious gift.

Shem (Hebrew) His name is God.

Sherborn (Old English) From the clear brook.

Sinclare (Old French) From Saint Clair.

Slevin (Gaelic) The mountain climber.

Sparky (Old English) Happy.

Stan (Old English) From the rocky meadow.

Stanislaus (Slavic) Stand of glory.

Steiner (Old German) Rock warrior.

Sterne (Old English) Austere.

Steward (Old English) The steward.

Stirling (Old English) Valuable.

Stuart (Old English) The steward.

Sumner (Old English) Church officer, summoner.

Tadeo (Latin) Praise.

Tamas (Hebrew) The devoted brother.

Tanner (Old English) Leatherworker, tanner.

Taro (Japanese) Firstborn male.

Terrell (Old German) Martial.

Theodore (Greek) Gift of God.

Thurston (Scandinavian) Thor's stone.

Trip (Old English) To dance or hop.

Tully (Gaelic) He who lives with God's peace.

Tylor (Old English) Maker of tiles.

Udell (Old English) Prosperous.

Ulric (Old German) Wolf-ruler.

Vance (Old English) Thresher.

Waring (Old German) Watchman or defender.

Waylon (Old English) From the land by the road.

Wendel (Old German) The wanderer.

Wilfrid (Old German) Resolute and peaceful.

Willy (Old German) Determined guardian.

Windham (Scottish Gaelic) Village near the winding road.

Wolfy (Old German) Wolf.

Xylon (Greek) From the forest.

Yardley (Old English) From the enclosed meadow.

Yehudi (Hebrew) Praise the Lord.

Yule (Old English) Born at Yuletide.

Yuriko (Hebrew) Jehovah is my light.

Zared (Hebrew) Ambush.

NUMBER NINE NAMES—UNISEX

Angie (Greek) Heavenly messenger.

Ariel (Hebrew) Lioness of God.

Aubrey (Old French) Blond ruler.

Bailey (Old French) Bailiff, steward.

Daniel (Hebrew) God is my judge.

Devin (Gaelic) Poet.

Essex (Old English) From the east.

Fortune (Latin) Fortunate.

Gabriel (Hebrew) Devoted to God.

Germaine (Old German) People of the spear.

Halley (Old English) Hay meadow.

Holly (Old English) Holly tree.

Jody (Hebrew) Admired.

Kacey (Gaelic) Vigilant.

Kip (Old English) From the pointed hill.

Lupe (Arabic) River of black stones.

Marlen (Old English) Falcon or hawk.

Maxy (Latin) Most excellent.

Morgen (Scottish Gaelic) From the edge of the sea.

Nikki (Greek) The people's victory.

Patsy (Latin) Wellborn, noble.

Regan (Gaelic) Little king.

Rocky (Old English) From the rock.

Ryann (Gaelic) Little king.

Ryon (Gaelic) Little king.

Sandy (Old English) From the sandy hill.

Shelly (Old English) From the meadow on the ledge.

Skyler (Dutch) Scholar.

Taran (Gaelic) Rocky pinnacle.

Tracey (Gaelic) Battler.

Vally (Old English) From the valley.

Vivien (Latin) Alive.

Appendix 1

MASTER LIST—FEMALE NAMES

Abigail (Hebrew) My father is joy. #5

Abra (Hebrew) Earth mother. #4

Abriana (Hebrew) Mother of the multitude. #1

Abrianna (33/6) (Hebrew) Mother of the multitude. #6

Acacia (Greek) Honorable. #9

Ada (Hebrew) Ornament. #6

Adabelle (Hebrew-French) Beautiful ornament. #6

Adalia (Old German) Noble. #1

Adalicia (Old German) Noble. #4

Adaline (Old German) Noble. #1

Adana (Hebrew) Of the red earth. #3

Adanna (Hebrew) Of the red earth. #8

Adara (Arabic) Virgin. #7

Addie (Old German) Of noble rank. #5

Adel (Old German) Of noble rank. #4

Adela (Old German) Of noble rank. #5

Adelaide (Old German) Of noble rank. #5

Adele (Old German) Of noble rank. #9

Adeline (Old German) Of noble rank. #5

Adelisa (Greek) Truthful. #6

Adelita (Old German) Of noble rank. #7

Adeliz (Greek) Truthful. #3

Adelle (Old German) Of noble rank. #3

Adena (Hebrew) Sensuous. #7

Adina (Hebrew) Ornament. #2

Adinah (Hebrew) Ornament. #1

Adira (Arabic) Strong. #6

Adolpha (Old German) Noble wolf. #3

Adonia (Greek) Beautiful. #8

Adora (Latin) Beloved. #3

Adorna (Latin) Adorned with jewels. #8

Adrianna (Latin) Black earth. #8

Adrienne (Latin) Black earth. #7

Afton (Old English) One from Afton. #2

Agatha (Greek) Good, kind. #2

Agnes (Greek) Pure. #1

Aida (Old English) Prosperous, happy. #6

Aidee (Greek) Well-behaved. #6

Ailani (Hawaiian) High chief. #1

Aileen (Greek) Light, a torch. #1

Aileene (33/6) (Greek) Light, a torch. #6

Aimee (Latin) The beloved. #6

Ainsley (Scottish Gaelic) From one's own meadow. #4

Aisha (African) Life. #2

Aisleen (Gaelic) Dream or vision. #2

Aislin (Gaelic) Dream or vision. #1

Akiko (Japanese) Light, bright. #2

Akilah (Arabic) Bright, smart. #6

Akima (Japanese) Autumn space. #8

Alair (Latin) Cheerful, glad. #5

Alameda (Spanish) Poplar tree. #1

Alana (11/2) (Gaelic) Comely, cheerful. #2

Alandra (Greek) Helper of humankind. #6

Alanna (Gaelic) Comely, cheerful. #7

Alanza (Old German) Noble and eager. #1

Alarice (Old German) Rules all. #4

Alberta (Old English) Noble and brilliant. #5

Albina (Latin) White. #3

Alcina (22/4) (Greek) Strong-minded. #4

Alda (Latin) Elder. #9

Aleda (Greek) Beautifully dressed. #5

Alejandra (Greek) Helper of humankind. #3

Alena (Greek) Light, a torch. #6

Aleria (Latin) Like an eagle. #1

Aletha (Greek) Truthful. #2

Alethea (Greek) Truthful. #7

Alexa (Greek) Helper of humankind. #7

Alexandra (Greek) Helper of humankind. #8

Alexia (Greek) Helper of humankind. #7

Alfonsine (Old German) Noble and eager. #5

Alfreda (Old English) Good counselor. #2

Alia (Israeli) Immigrant to a new home. #5

Alice (Greek) Truthful. #3

Alicia (Greek) Truthful. #8

Alida (Greek) Beautifully dressed. #9

Alika (Greek) Helper of humankind. #7

Alima (Arabic) Learned in music and dance. #9

Alina (Slavic) Bright, beautiful. #1

Aline (Old German) Of noble rank. #5

Alisa (Greek) Truthful. #6

Alisabeth (Hebrew) God's promise. #5

Alise (Greek) Truthful. #1

Alisha (Greek) Truthful. #5

Alison (Greek) Truthful. #7

Alissa (Greek) Truthful. #7

Alita (Old German) Of noble rank. #7

Alix (Greek) Helper of humankind. #1

Aliza (22/4) (Greek) Truthful. #4

Alizah (Hebrew) Joyful. #3

Allaryce (Old German) Rules all. #5

Allegra (Latin) Exuberantly cheerful. #2

Allena (Greek) Light, a torch. #9

Allison (Greek) Truthful. #1

Alma (Latin) Soul, spirit. #9

Almeda (Latin) Pressing toward the goal. #9

Almira (Arabic) Exalted. #9

Aloha (Hawaiian) Love. #1

Aloise (Old German) Famous in war. #7

Aloma (Spanish) Dove. #6

Alondra (Spanish) Lark. #2

Alpha (Greek) First one. #2

Alphonsia (Old German) Noble and eager. #5

Alta (Latin) High or lofty. #7

Altagracia (Spanish) Divine grace. #1

Althea (Greek) Wholesome healing. #2

Alula (11/2) (Arabic) The first. #2

Alura (Old English) Divine counselor. #8

Alva (Latin) White, fair. #9

Alvern (Latin) Spring, greening. #9

Alvina (Old German) Beloved by all. #5

Alwyn (Old German) Beloved by all. #3

Alyce (Greek) Truthful. #1

Alyda (Greek) Beautifully dressed. #7

Alyssa (Greek) Truthful. #5

Ama (Ghanese) Saturday's child. #6

Amabel (Latin) Lovable. #7

Amabelle (Latin) Lovable. #6

Amala (Arabic) Beloved. #1

Amalia (Old German) Industrious. #1

Amanda (Latin) Worthy of being loved. #7

Amantha (22/4) (Greek) Beyond death. #4

Amapola (Arabic) Poppy. #5

Amara (Greek) Of eternal beauty. #7

Amarantha (Greek) Beyond death. #5

Amargo (Greek) Of eternal beauty. #1

Amaryllis (Greek) Amaryllis flower. #2

Amaya (Japanese) Night rain. #5

Amber (Old French) Amber jewel. #3

Ambrosine (Greek) Immortal. #6

Ameera (Arabic) Princess. #7

Amelia (Old German) Industrious. #5

Amelinda (Spanish) Beloved and pretty. #5

Amethyst (Greek) Against intoxication. #3

Amie (Latin) The beloved. #1

Amina (Arabic) Trustworthy. #2

Aminta (22/4) (Greek) Protector. #4

Amira (Arabic) Princess. #6

Amita (Israeli) Truth. #8

Amity (Latin) Friendship. #5

Amy (Latin) The beloved. #3

Amye (Latin) The beloved. #8

Ana (Hebrew) Full of grace. #7

Anabel (Hebrew-Latin) Graceful, beautiful. #8

Anabella (Hebrew-Latin) Graceful, beautiful. #3

Anabelle (Hebrew-Latin) Graceful, beautiful. #7

Analeigh (Hebrew-Old English) Graceful, meadow. #3

Analise (Hebrew) Graced with God's promise. #7

Anamaria (Hebrew) Bitter, graceful. #4

Anamarie (Hebrew) Bitter, graceful. #8

Anarosa (Hebrew-Greek) Graceful rose. #6

Anastacia (Greek) Of the resurrection. #6

Anastasia (22/4) (Greek) Of the resurrection. #4

Anatola (Greek) From the east. #1

Anda (11/2) (Japanese) Meet at the field. #2

Andee (Latin) Womanly. #2

Andrea (Latin) Womanly. #7

Aneko (Japanese) Older sister. #1

Angela (22/4) (Greek) Heavenly messenger. #4

Angelica (Greek) Heavenly messenger. #7

Angelina (Greek) Heavenly messenger. #9

Angeline (Greek) Heavenly messenger. #4

Angelique (Latin) Like an angel. #1

Angharad (Welsh) Free from shame. #9

Anissa (Hebrew) Full of grace. #9

Anita (Hebrew) Full of grace. #9

Anjanette (Hebrew) God's gift of grace. #9

Ann (11/2) (Hebrew) Full of grace. #2

Anna (Hebrew) Full of grace. #3

Annabel (22/4) (Hebrew-Latin) Graceful, beautiful. #4

Annabelle (Hebrew-Latin) Graceful, beautiful. #3

Annalisa (Hebrew) Graced with God's promise. #8

Annamaria (Hebrew) Bitter, graceful. #9

Anne (Hebrew) Full of grace. #7

Annelise (Hebrew) Graced with God's promise. #7

Annemarie (Hebrew) Bitter, graceful. #8

Annette (Hebrew) Full of grace. #7

Annie (Hebrew) Full of grace. #7

Annis (Hebrew) Full of grace. #3

Annora (Old English) Light and graceful. #9

Anona (Latin) Yearly crops. #9

Anora (22/4) (Old English) Light and graceful. #4

Anthea (22/4) (Greek) Like a flower. #4

Antoinette (Latin) Praiseworthy, without a peer. #6

Antonia (Latin) Praiseworthy, without a peer. #2

Antonis (Latin) Praiseworthy, without a peer. #2

Anya (Hebrew) Full of grace. #5

Aphra (Hebrew) Dust. #8

Aphrodite (Greek) Goddess of love and beauty. #6   •

Apollonia (Latin) Belonging to Apollo. #5

April (Latin) Forthcoming. #2

Aquilina (Spanish) An eagle. #3

Arabella (Latin) Beautiful altar. #7

Arabelle (Latin) Beautiful altar. #2

Araceli (Latin) Altar of heaven. #4

Aracellie (Latin) Altar of heaven. #3

Aracelly (Latin) Altar of heaven. #5

Aracely (Latin) Altar of heaven. #2

Arcadia (Greek) Pastoral simplicity and happiness. #1

Ardis (Hebrew) Flowering field. #6

Ardith (33/6) (Hebrew) Flowering field. #6

Areta (Greek) Best. #9

Aretha (Greek) Best. #8

Argenta (Latin) Silvery. #3

Ariadne (Greek) Holy one. #7

Ariana (Greek) Holy one. #8

Arleigh (Old German) Meadow of the hare. #6

Arlene (Gaelic) A pledge. #1

Arliss (Old German) Womanly strength. #6

Armanda (Old German) Army woman. #7

Armida (Latin) Small warrior. #1

Arsenia (Greek) Potent. #4

Arthurine (Celtic-Welsh) Noble, bear-heroine. #6

Artimis (Greek) Divine moon goddess. #8

Asha (11/2) (African) Life. #2

Ashlynn (Old English) From the ash tree meadow. #3

Ashtin (Old English) From the ash tree town. #8.

Astra (Latin) Star. #5

Astrid (Old Norse) Divine strength. #8

Asucena (Arabic) Lily. #1

Atalanta (Greek) Mighty adversary. #7

Athena (22/4) (Greek) Wisdom. #4

Athene (Greek) Wisdom. #8

Auberta (Old English) Noble and brilliant. #5
Audrey (Old English) Noble strength. #2
Augusta (Latin) Majestic dignity. #9
Aurelia (Latin) Golden. #4
Aurora (Latin) Daybreak. #2
Autumn (Latin) The fall season. #9
Ava (Latin) Birdlike. #6
Averil (Latin) Forthcoming. #4
Avis (Old English) Refuge in battle. #6
Aviva (Hebrew) Springtime. #1
Avril (Latin) Forthcoming. #8
Aya (Japanese) Patterned cloth, damask. #9
Ayla (Hebrew) Oak tree. #3
Azalea (Latin) Dry, azalea flower. #1
Aziza (Arabic) Beloved. #9
Azucena (Arabic) Lily. #8
Azura (22/4) (Old French) Sky blue. #4
Azure (Old French) Sky blue. #8
Azusa (Arabic) Lily. #5
Azusena (Arabic) Lily. #6
Bahira (Arabic) Dazzling. #3
Bambi (Italian) Child. #9
Baptista (Latin) Free from sin. #7
Barbara (Latin) Beautiful stranger. #7

Barbie (Latin) Beautiful stranger. #1
Barbra (Latin) Beautiful stranger. #6
Basilia (Latin) Queenly, regal. #8
Bathsheba (Hebrew) Daughter of the oath. #3
Bea (Latin) Bringer of joy. #8
Beata (11/2) (Latin) Happy. #2
Beatrice (Latin) Bringer of joy. #9
Beatrix (Latin) Bringer of joy. #7
Beatriz (Latin) Bringer of joy. #9
Becca (Hebrew) The captivator. #5
Beckie (Hebrew) The captivator. #8
Becky (Hebrew) The captivator. #1
Beda (Old English) Warrior maiden. #3
Bee (Latin) Bringer of joy. #3
Bekky (Hebrew) The captivator. #9
Belen (Spanish) From Bethlehem. #2
Belinda (Spanish) Pretty. #2
Bella (French) Beautiful. #5
Belle (French) Beautiful. #9
Benecia (Latin) Blessed. #3
Benedicta (Latin) Blessed. #9
Benicia (Latin) Blessed. #7

Benigna (Latin) A great lady. #7

Benita (Latin) Blessed. #6

Berenice (Greek) Bringer of victory. #7

Bernadette (Old German) Brave as a bear. #4

Bernadine (Old German) Brave as a bear. #9

Bernice (Greek) Bringer of victory. #2

Berta (Old English) Bright. #1

Bertha (Old German) Shining. #9

Bertrade (Old English) Shining advisor. #1

Beryl (Greek) The sea-green jewel. #8

Bess (Hebrew) God's promise. #9

Besse (Hebrew) God's promise. #5

Bessie (Hebrew) God's promise. #5

Bessy (Hebrew) God's promise. #7

Beth (Hebrew) House of God. #8

Bethany (Aramaic) House of poverty. #3

Bethea (Hebrew) Maidservant of Jehovah. #5

Betina (Hebrew) God's promise. #6

Betsy (Hebrew) God's promise. #8

Bette (Hebrew) God's promise. #7

Betty (Hebrew) God's promise. #9

Beulah (22/4) (Hebrew) Married. #4

Beverley (Old English) From the beaver meadow. #4

Bianca (Latin) White. #3

Bibi (22/4) (Latin) Alive. #4

Bina (Hebrew) Understanding. #8

Binah (Hebrew) Understanding. #7

Birdie (English) Birdlike. #2

Blanca (Spanish) White, fair. #6

Blanche (Old French) White, fair. #9

Bliss (Old English) Joy, bliss. #7

Blithe (Old English) Joyful. #2

Blodwyn (Welsh) White flower. #5

Blondelle (French) Little fair one. #9

Blondie (French) Little fair one. #7

Blossom (Old English) Flowerlike. #5

Blyth (22/4) (Old English) Joyful. #4

Blythe (Old English) Joyful. #9

Bonita (Spanish) Pretty. #7

Bonnie (Scottish-English) Beautiful, pretty. #5

Bonny (Scottish-English) Beautiful, pretty. #7

Branwyn (Old Welsh) White raven. #7

Breezy (English) Lightly windy. #9

Brena (22/4) (Gaelic) Little raven. #4

Brenda (Old English) Firebrand. #8

Brenna (Gaelic) Little raven. #9

Briana (Celtic) Strength with virtue and honor. #9

Brianna (Celtic) Strength with virtue and honor. #5

Brianne (Celtic) Strength with virtue and honor. #9

Briar (Old English) A thorny plant. #3

Bridget (Celtic) Strong and mighty. #2

Bridie (Celtic) Strong and mighty. #2

Brier (Old English) A thorny plant. #7

Brigitta (Celtic) Strong and mighty. #5

Brigitte (Celtic) Strong and mighty. #9

Brina (Old Norse) Protector. #8

Briona (Celtic) Strength with virtue and honor. #5

Briony (Celtic) Strength with virtue and honor. #2

Brita (Latin) From England. #5

Brittania (Latin) From England. #4

Brittany (Latin) From England. #1

Britteny (Latin) From England. #5

Brittney (Latin) From England. #5

Bronwen (Old Welsh) White-bosomed. #1

Bronwyn (Old Welsh) White-bosomed. #3

Brunella (Old German) Brown-haired. #4

Brunetta (Old German) Brown-haired. #2

Brunhilda (Old German) Armed battle maid. #8

Brunhilde (Old German) Armed battle maid. #3

Bryanna (Celtic) Strength with virtue and honor. #3

Brydie (Celtic) Strong and mighty. #9

Bryna (Celtic) Strength with virtue and honor. #6

Brynn (Celtic) The heights. #1

Bryony (Celtic) Strength with virtue and honor. #9

Bunny (22/4) (English) Little rabbit. #4

Burgundy (French) A reddish-purple color. #4

Cadence (Greek) Pastoral simplicity and happiness. #8

Cady (Greek) Pastoral simplicity and happiness. #6

Cailin (Gaelic) Thin, slender. #3

Caitlin (Greek) Pure maiden. #5

Caitlyn (Greek) Pure maiden. #3

Calandra (Greek) Lark. #9

Caledonia (Latin) Scottish lassie. #1

Calida (Spanish) Warm, ardent. #3

Calista (Greek) Most beautiful woman. #2

Calla (11/2) (Greek) Beautiful. #2

Callan (Gaelic) Powerful in battle. #7

Callie (Gaelic) Thin, slender. #6

Calliope (Greek) Beautiful voice. #1

Calvinna (Latin) Bald. #4

Calypso (Greek) Concealer. #1

Cameo (Italian) Sculptured jewel. #1

Camilla (Latin) Young ceremonial attendant. #6

Camille (Latin) Young ceremonial attendant. #1

Candace (22/4) (Greek) Glittering, glowing white. #4

Candice (Greek) Glittering, glowing white. #3

Candida (Greek) Glittering, glowing white. #9

Candide (Greek) Glittering, glowing white. #4

Candra (Latin) Luminescent. #5

Candy (Greek) Glittering, glowing white. #2

Caprice (Italian) Fanciful. #1

Cara (Latin) Dear. #5

Caresse (French) Beloved. #7

Cari (22/4) (Turkish) Flows like water. #4

Carina (Latin) Dear. #1

Carissa (Greek) Loving. #7

Carita (Latin) Charity. #7

Carla (Old German) Little woman born to command. #8

Carlie (Old German) Little woman born to command. #3

Carlotta (Old German) Little woman born to command. #9

Carly (Old German) Little woman born to command. #5

Carmela (Hebrew) God's vineyard. #8

Carmelita (Hebrew) God's vineyard. #1

Carmen (Latin) Song. #9

Caro (Old German) Little woman born to command. #1

Carole (Old German) Noble and strong. #9

Carolina (Old German) Little woman born to command. #1

Caroline (Old German) Little woman born to command. #5

Carolyn (Old German) Little woman born to command. #7

Carrie (Old German) Little woman born to command. #9

Casandra (Greek) Disbelieved by men. #7

Cassandra (Greek) Disbelieved by men. #8

Cassia (Greek) Sweet-scented spice. #7

Cassie (Greek) Disbelieved by men. #2

Catharine (Greek) Pure maiden. #7

Catherine (Greek) Pure maiden. #2

Cathy (Greek) Pure maiden. #3

Catriona (Greek) Pure maiden. #9

Cayley (Greek) Pure maiden. #8

Ceara (Gaelic) Spear. #1

Cecilia (33/6) (Latin) Dim-sighted or blind. #6

Cecily (Latin) Dim-sighted or blind. #3

Cedra (22/4) (Old English) Battle chieftain. #4

Cedrica (Old English) Battle chieftain. #7

Ceiridwen (Welsh) Goddess of bards. #9

Celeste (Latin) Heavenly. #6

Celestine (Latin) Heavenly. #2

Celia (Latin) Dim-sighted or blind. #3

Celina (Greek) Moon. #8

Ceres (Latin) Of the spring. #5

Cerise (French) Cherry, cherry-red. #5

Chandel (Old French) Candle. #2

Chandelle (Old French) Candle. #1

Chandra (Sanskrit) Moonlike. #4

Chanel (French) Channel. #7

Chantal (Latin) Song. #5

Chantalle (Latin) Song. #4

Chantel (Latin) Song. #9

Chantilly (French) Lace from Chantilly. #5

Charis (Greek) Grace. #4

Charity (Latin) Brotherly love. #3

Charleen (French) Little and womanly. #3

Charlene (French) Little and womanly. #3

Charlotte (French) Little and womanly. #3

Charmaine (Latin) Song. #9

Charo (Spanish) Rosary. #9

Chastity (33/6) (Latin) Purity. #6

Chelsea (Old English) A port of ships. #8

Chelsey (Old English) A port of ships. #5

Chelsie (Old English) A port of ships. #7

Cher (Old French) Beloved. #7

Cherie (Old French) Beloved. #3

Cherry (English) Cherry fruit. #5

Cheryl (Old French) Little and womanly. #8

Chessa (Slavic) At peace. #1

Chica (Spanish) Small girl. #6

Chiquita (Spanish) Little one. #7

Chloe (Greek) Young grass. #7

Chloris (Greek) Pale. #3

Chrissy (Greek) Anointed. #2

Christa (33/6) (Greek) Anointed. #6

Christabel (Greek) Beautiful anointed one. #7

Christiana (Greek) Anointed. #3

Christina (Greek) Anointed. #2

Christine (Greek) Anointed. #6

Christy (Greek) Anointed. #3

Ciana (Hebrew) God's gracious gift. #1

Ciandra (Hebrew) God's gracious gift. #5

Ciara (Gaelic) Little and dark-skinned. #5

Cicely (Latin) Dim-sighted or blind. #3

Cinderella (French) Little one of the ashes. #2

Cindi (Greek) The moon. #3

Cindy (Greek) The moon. #1

Cinnamon (English) Cinnamon spice. #2

Claire (Latin) Bright, shining girl. #3

Clara (Latin) Bright, shining girl. #8

Clarabella (French) Brilliant, beautiful. #4

Clarabelle (French) Brilliant, beautiful. #8

Clarebelle (French) Brilliant, beautiful. #3

Claribel (French) Brilliant, beautiful. #8

Claribelle (French) Brilliant, beautiful. #7

Clarice (33/6) (Latin-Greek) Most brilliant. #6

Clarissa (Latin-Greek) Most brilliant. #1

Claudia (Latin) Lame. #6

Claudine (33/6) (Latin) Lame. #6

Clea (Greek) Famous. #3

Clemence (33/6) (Latin) Merciful. #6

Clementina (Latin) Merciful. #6

Clementine (Latin) Merciful. #1

Cleo (Greek) Famous. #8

Cleopatra (Greek) Famous. #1

Cleta (Greek) Summoned. #5

Clio (Greek) Famous. #3

Clorinda (Latin) Famed for her beauty. #4

Cloris (Greek) Pale. #4

Clotilda (old German) Famous battle maiden. #4

Clover (Old English) Clover blossom. #3

Coleen (Gaelic) Girl. #9

Colette (Greek) The people's victory. #8

Colleen (Gaelic) Girl. #3

Columba (22/4) (Latin) Dovelike, gentle. #4

Columbia (Latin) Dovelike, gentle. #4

Conception (Latin) Conception. #6

Concetta (Latin) Conception. #9

Conchita (Latin) Conception. #1

Concordia (Latin) Harmony. #1

Consolata (Latin) One who consoles. #1

Constance (Latin) Firmness, constancy. #4

Constancia (Latin) Firmness, constancy. #9

Consuela (Spanish) Consolation. #9

Consuelo (Spanish) Consolation. #5

Cookie (English) Cookie. #4

Cora (Greek) The maiden. #1

Corabella (33/6) (Greek-French) Beautiful maiden. #6

Coral (22/4) (Latin) Small sea creature. #4

Corazon (Spanish) Heart. #2

Cordelia (Welsh) Jewel of the sea. #4

Coretta (Greek) The maiden. #1

Corina (33/6) (Greek) the maiden. #6

Corine (Greek) The maiden. #1

Corinna (Greek) The maiden. #2

Corinne (Greek) The maiden. #6

Corinthia (Greek) Woman of Corinth. #7

Corissa (Greek) The maiden. #3

Corliss (Old English) Cheerful, good-hearted. #5

Cornelia (Latin) Yellow, horn-colored. #5

Cosette (French) Little pet. #6

Cosina (Greek) World harmony. #7

Cozette (French) Little pet. #4

Crissi (Greek) Anointed. #5

Crissy (Greek) Anointed. #3

Cristina (Greek) Anointed. #3

Cristobel (Greek) Beautiful anointed one. #4

Crystal (Latin) Without deception. #8

Cybele (Greek) A prophetess. #7

Cybil (Greek) A prophetess. #6

Cybill (Greek) A prophetess. #9

Cynthera (Greek) Goddess of love and beauty. #4

Cynthia (Greek) The moon. #8

Cyrilla (Greek) Mistress, ruler. #8

Dacey (Gaelic) Southerner. #2

Dacia (Greek) From Dacia. #9

Dacy (Gaelic) Southerner. #6

Dagmar (Old German) Glorious day. #8

Dahlia (Norse) From the valley. #8

Daisy (22/4) (Old English) Eye-of-the-day. #4

Damara (Greek) Gentle. #2

Damaris (Greek) Gentle. #2

Damita (Spanish) Little noble lady. #3

Danica (Slavic) The morning star. #5

Daniela (Hebrew) God is my judge. #1

Danielle (Hebrew) God is my judge. #8

Danika (22/4) (Slavic) The morning star. #4

Daphne (Greek) Laurel or bay tree. #3

Dara (Hebrew) Compassion. #6

Daria (Greek) Queenly. #6

Darlene (Old French) Little darling. #5

Davina (Hebrew) Beloved. #6

Davita (Hebrew) Beloved. #3

Dawn (Old English) The break of day. #6

Deana (Latin) Divine moon goddess. #7

Deanna (Latin) Divine moon goddess. #3

Debbie (Hebrew) The bee. #9

Debby (Hebrew) The bee. #2

Debi (Hebrew) The bee. #2

Deborah (Hebrew) The bee. #8

Debra (Hebrew) The bee. #3

Dee (Welsh) Black, dark. #5

Deeann (Latin) Divine moon goddess. #7

Deidre (Gaelic) Sorrow. #9

Deirdre (Gaelic) Sorrow. #9

Deitra (Greek) From the fertile land. #3

Deka (Somali) Pleasing. #3

Delcine (Latin) Sweet and charming. #7

Delfina (33/6) (Greek) Calmness. #6

Delia (22/4) (Greek) Visible. #4

Delicia (Latin) Gives pleasure. #7

Delilah (33/6) (Hebrew) Delicate, languishing, amorous. #6

Della (Greek) Visible. #7

Delphine (Greek) Calmness. #1

Delta (Greek) Fourth-born. #6

Demelza (English) From Demelza in Cornwall. #3

Demeter (Greek) From the fertile land. #7

Demetra (Greek) From the fertile land. #3

Demetria (Greek) From the fertile land. #3

Demi (22/4) (Latin) Half, partly belonging. #4

Dena (Latin) Divine moon goddess. #6

Denice (Greek) Wine goddess. #4

Denise (Greek) Wine goddess. #2

Desire (33/6) (French) Longed for. #6

Desiree (French) Longed for. #2

Desma (Greek) A pledge. #6

Dessa (Greek) Wandering. #3

Destinee (Old French) Destiny. #9

Destiny (33/6) (Old French) Destiny. #6

Deva (Sanskrit) Divine. #5

Devany (Old English) Defender of Devonshire. #8

Devi (22/4) (Sanskrit) Divine. #4

Devon (Old English) Defender of Devonshire. #6

Devona (Old English) Defender of Devonshire. #7

Devondra (Old English) Defender of Devonshire. #2

Devonna (Old English) Defender of Devonshire. #3

Devonne (Old English) Defender of Devonshire. #7

Devony (Old English) Defender of Devonshire. #4

Devora (Hebrew) The bee. #2

Dextra (Latin) Skillful, adept. #9

Diahann (33/6) (Latin) Divine moon goddess. #6

Diana (Latin) Divine moon goddess. #2

Diandre (Latin) Divine moon goddess. #1

Diane (Latin) Divine moon goddess. #6

Dianna (Latin) Divine moon goddess. #7

Diantha (Latin-Greek) Divine moon goddess. #3

Dianthe (Latin-Greek) Divine moon goddess. #7

Dilys (Welsh) Perfection. #6

Dina (Hebrew) Judged. #1

Dinah (Hebrew) Judged. #9

Dinora (Hebrew) Judged. #7

Dinorah (Hebrew) Judged. #6

Dione (Greek) Daughter of heaven and earth. #2

Dionne (Greek) Daughter of heaven and earth. #7

Divina (Latin) Divine. #5

Dixie (33/6) (American) Girl of the south. #6

Diza (22/4) (Hebrew) Joy. #4

Docila (Latin) Gentle teacher. #8

Dodie (Hebrew) Beloved. #1

Dody (Hebrew) Beloved. #3

Dolly (Greek) Gift of God. #5

Dolores (Spanish) Sorrows. #7

Domina (Latin) A noble lady. #2

Dominga (Latin) The Lord's day. #9

Dominica (Latin) Belonging to the Lord. #5

Dominique (Latin) Belonging to the Lord. #8

Donata (Latin) The gift. #1

Donelle (Latin) Lady. #4

Donna (Latin) Lady. #3

Dora (Greek) Gift. #2

Dorcas (Greek) Grace of the gazelle. #6

Doreen (Gaelic) Sullen. #7

Dori (Greek) Gift. #1

Doria (Greek) Gift. #2

Dorinda (Greek) Beautiful golden gift. #2

Doris (Greek) From the ocean. #2

Dorothea (Greek) Gift of God. #5

Dorothy (Greek) Gift of God. #6

Dorrie (Greek) Gift. #6

Dory (Greek) Gift. #8

Dot (Greek) Gift of God. #3

Dottie (Greek) Gift of God. #1

Dotty (Greek) Gift of God. #3

Drucilla (Latin) Strong. #8

Drusilla (33/6) (Latin) Strong. #6

Duana (Gaelic) Dark. #5

Dulce (Latin) Sweet and charming. #9

Dulcea (Latin) Sweet and charming. #1

Dulcie (Latin) Sweet and charming. #9

Dulcine (Latin) Sweet and charming. #5

Dyane (22/4) (Latin) Divine moon goddess. #4

Earline (Old English) Noblewoman. #1

Eartha (Old English) Of the earth. #8

Easter (Old English) Easter time. #5

Ebony (Greek) A hard, dark wood. #7

Echo (22/4) (Greek) Sound. #4

Eda (Old German) Strives. #1

Edana (Celtic) Flame. #7

Edda (Old German) Strives. #5

Edie (Old English) Rich gift. #5

Edina (Old English) Rich friend. #6

Edith (Old English) Rich gift. #1

Edmonda (Old English) Prosperous protector. #2

Edna (Hebrew) Rejuvenation. #6

Edwina (Old English) Rich friend. #2

Edythe (Old English) Rich gift. #4

Effie (Greek) Well spoken of. #4

Egypt (English) From Egypt. #1

Eileen (Greek) Light, a torch. #5

Eirene (Greek) Goddess of peace. #2

Eirlys (Old Welsh) Snow drop. #7

Elaine (Greek) Light, a torch. #1

Elana (Hebrew) Tree. #6

Elberta (Old English) Noble and brilliant. #9

Eldora (Greek) Gift of the sun. #1

Eleanor (Old French) Light. #7

Electra (Greek) Brilliant, shining. #1

Elena (Greek) Light, a torch. #1

Eleni (Greek) Light, a torch. #9

Elfreda (33/6) (Old English) Elfin magic. #6

Elfrida (Old English) Elfin magic. #1

Elga (Slavic) Consecrated. #7

Eliana (Greek) Light, a torch. #6

Elinor (Old French) Light. #1

Elisabeth (Hebrew) God's promise. #9

Elise (Hebrew) God's promise. #5

Elisha (Hebrew) God is salvation. #9

Elissa (Hebrew) God's promise. #2

Elita (Latin) Chosen. #2

Eliza (Hebrew) God's promise. #8

Elizabeth (Hebrew) God's promise. #7

Elke (Old German) Of noble rank. #6

Ella (Greek) Light. #3

Ellen (Greek) Light. #3

Ellie (Greek) Light. #7

Elly (Greek) Light. #9

Elma (Old German) God's protection. #4

Elna (Greek) Light, a torch. #5

Elnora (Old French) Light. #2

Eloisa (Old German) Famous woman warrior. #7

Eloise (Old German) Famous woman warrior. #2

Elrica (Old German) Ruler of all. #3

Elsa (Hebrew) God's promise. #1

Elsbeth (Hebrew) God's promise. #8

Elsie (Hebrew) God's promise. #5

Elspeth (Hebrew) God's promise. #4

Elva (Old English) Elfin, good. #4

Elvia (22/4) (Old English) Elfin, good. #4

Elvina (Old English) Elf-friend. #9

Elvira (Latin) White, blond. #4

Elyse (Hebrew) God's promise. #3

Elysia (Latin) Sweetly blissful. #8

Emanuele (Hebrew) God is with us. #4

Emerald (French) Emerald jewel. #4

Emilia (Latin) Flattering. #4

Emily (Latin) Flattering. #1

Emma (Old German) One who leads the universe. #5

Emmaline (Old German) One who leads the universe. #9

Emmeline (Old German) One who leads the universe. #4

Emogene (Latin) Image. #1

Ena (11/2) (Gaelic) Little ardent one. #2

Enid (Old Welsh) Purity. #5

Enrica (Old German) Lord of the manor. #5

Enyd (Old Welsh) Purity. #3

Erica (Old Norse) Ever-powerful, ever-ruler. #9

Erika (Old Norse) Ever-powerful, ever-ruler. #8

Eriko (Japanese) Child with a collar. #4

Erma (Latin) High-ranking person. #1

Erna (Old English) Eagle. #2

Ernestina (Old English) Earnest. #6

Ernestine (Old English) Earnest. #1

Esme (Old French) Esteemed. #6

Esmeralda (33/6) (Spanish) Emerald. #6

Esperanza (Spanish) Hope. #6

Estefania (Greek) Crowned. #8

Estefany (Greek) Crowned. #5

Estella (Persian) A star. #2

Estelle (Persian) A star. #6

Esther (Persian) A star. #3

Ethel (Old English) Noble. #5

Ethyl (Old English) Noble. #7

Etta (Old German) Little. #1

Eudora (Greek) Honored gift. #1

Eugenia (Greek) Wellborn noble. #8

Eulalia (Greek) Sweet-spoken. #7

Eulalie (Greek) Sweet-spoken. #2

Eunice (Greek) The people's victory. #3

Euphemia (Greek) Auspicious speech. #6

Eustacia (Greek) Steadfast. #7

Eva (Hebrew) Life-giving. #1

Evangelina (Greek) Bearer of good tidings. #9

Evangeline (Greek) Bearer of good tidings. #4

Evania (Greek) Tranquil. #7

Eve (Hebrew) Life-giving. #5

Evelina (Hebrew) Life-giving. #5

Eveline (Hebrew) Life-giving. #9

Evita (Hebrew) Life-giving. #3

Evonne (Hebrew) Life-giving. #3

Fabia (Latin) Bean grower. #1

Fabiola (Latin) Bean grower. #1

Fae (Old German) Belief in God. #3

Faith (Old German) Belief in God. #8

Fallon (Gaelic) Grandchild of the ruler. #6

Fanella (Gaelic) White-shouldered. #6

Fannie (Latin) Free. #4

Fanny (Latin) Free. #6

Farah (Old English) Beautiful, pleasant. #7

Farrah (Old English) Beautiful, pleasant. #7

Fatima (Arabic) Wean or abstain. #5

Fatimah (Arabic) Wean or abstain. #4

Faustine (Latin) Fortunate. #5

Fawn (Old French) Young deer. #8

Fawna (Old French) Young deer. #9

Fawne (22/4) (Old French) Young deer. #4

Fay (Old German) Belief in God. #5

Faye (Old German) Belief in God. #1

Fedora (Greek) Gift of God. #4

Felda (Old German) From the field. #1

Felice (Latin) Joyous, fortunate. #4

Felicia (Latin) Joyous, fortunate. #9

Felicidad (Latin) Joyous, fortunate. #8

Felicie (Latin) Joyous, fortunate. #4

Felicitas (Latin) Joyous, fortunate. #3

Felicity (Latin) Joyous, fortunate. #8

Felipa (Greek) Lover of horses. #4

Felisha (33/6) (Latin) Joyous, fortunate. #6

Fenella (Gaelic) White-shouldered. #1

Feodora (Greek) Gift of gold. #1

Fern (Old English) A fern. #7

Fernanda (Old German) World-daring. #9

Fiala (Czech) Violet. #2

Fiammetta (Italian) A flickering fire. #7

Fidelia (Latin) Faithful. #1

Fidelity (Latin) Faithful. #9

Fidelma (Latin-Hebrew) Faithful, bitter. #5

Fifi (Hebrew) He shall add. #3

Filma (Old English) Misty veil. #5

Filomena (Greek) Love song. #3

Finella (Gaelic) White-shouldered. #5

Fiona (Gaelic) Fair. #9

Fionn (Gaelic) Fair. #4

Fionna (Gaelic) Fair. #5

Fionnula (Gaelic) White-shouldered. #2

Flair (French-English) Style, verve. #1

Flavia (Latin) Yellow-haired. #6

Fleur (Latin) Flower. #8

Flor (Latin) Flower. #6

Flora (Latin) Flowering or blooming. #7

Florencia (Latin) Flowering or blooming. #2

Florentina (Latin) Flowering or blooming. #6

Florida (Latin) Flowering or blooming. #2

Florimel (French-Greek) Dark flower. #9

Florina (Latin) Flowering or blooming. #3

Florinda (Latin) Flowering or blooming. #7

Florrie (Latin) Flowering or blooming. #2

Flossie (Latin) Flowering or blooming. #4

Flower (Latin) Flower. #7

Fontana (Old French) Fountain, spring. #8

Fontanne (Old French) Fountain, spring. #8

Fortuna (Latin) Fortunate. #5

Frances (Latin) Free. #3

Francesca (Latin) Free. #7

Francine (Latin) Free. #7

Francisca (Latin) Free. #2

Francoise (Latin) Free. #9

Frannie (Latin) Free. #4

Franny (33/6) (Latin) Free. #6

Frayda (Old German) Peaceful ruler. #1

Freda (Old German) Peaceful ruler. #7

Frederica (Old German) Peaceful ruler. #6

Fredericka (Old German) Peaceful ruler. #8

Freya (Old Norse) Noble lady. #1

Frieda (Old German) Peaceful ruler. #7

Fritzi (Old German) Peaceful ruler. #7

Gabriela (Hebrew) Devoted to God. #1

Gabriele (Hebrew) Devoted to God. #5

Gabrielle (Hebrew) Devoted to God. #8

Gaea (Greek) The earth. #5

Gaia (Greek) The earth. #9

Gala (Old English) Gay, lively. #3

Galina (Greek) Calm. #8

Garda (22/4) (Old German) The protected. #4

Gardenia (Scottish) Gardenia flower. #5

Gay (Old French) Bright and lively. #6

Gaye (Old French) Bright and lively. #2

Gayla (Old French) Bright and lively. #1

Gayle (Old French) Bright and lively. #5

Gem (Latin) Precious stone. #7

Gemma (Latin) Precious stone. #3

Gena (French) Pure white wave. #9

Geneva (French) Pure white wave. #9

Genevieve (French) Pure white wave. #4

Genevra (French) Pure white wave. #9

Genifer (French) Pure white wave. #1

Genna (French) Pure white wave. #5

Georgeanne (Greek) Landholder, farmer. #1

Georgette (Greek) Landholder, farmer. #3

Georgia (Greek) Landholder, farmer. #8

Georgiana (Greek) Landholder, farmer. #5

Georgina (Greek) Landholder, farmer. #4

Georgine (Greek) Landholder, farmer. #8

Geovanna (Hebrew) God's gracious gift. #7

Geralda (Old German) Spear-mighty. #3

Geraldine (Old German) Spear-mighty. #3

Germana (Old German) People of the spear. #5

Gertrude (Old German) Spear-strength. #8

Gianna (Hebrew) God's gracious gift. #1

Gigi (Old German) Brilliant hostage. #5

Gilberte (Old German) Brilliant hostage. #6

Gilda (Old English) Covered with gold. #6

Gillian (Greek) Young in heart and mind. #1

Gina (22/4) (Latin) Queen. #4

Ginamarie (Latin-Hebrew) Queen, bitter. #5

Ginger (Latin) Ginger flower or ginger spice. #6

Ginny (33/6) (Latin) Virginal. #6

Giordana (Hebrew) Down-flowing. #6

Giovanna (Hebrew) God is gracious. #2

Gipsy (Old English) Wanderer. #4

Gisela (Old German) Pledge or hostage. #8

Giselle (33/6) (Old German) Pledge or hostage. #6

Gitana (Old English) Wanderer. #7

Giuliana (Greek) Young in heart and mind. #2

Gladys (Celtic) Princess. #5

Glenda (Old Welsh) Dweller in a valley. #7

Glenna (Old Welsh) Dweller in a valley. #8

Glennis (Old Welsh) Dweller in a valley. #8

Gloria (Latin) Glory. #8

Gloriana (Latin) Glory. #5

Gloriane (Latin) Glory. #9

Glory (Latin) Glory. #5

Glynis (Old Welsh) Dweller in a valley. #5

Glynnis (Old Welsh) Dweller in a valley. #1

Godiva (Old English) Gift of God. #4

Golda (Old English) Golden. #3

Goldie (Old English) Golden. #7

Goldy (Old English) Golden. #9

Grace (Latin) Graceful. #7

Gracia (Latin) Graceful. #3

Gracie (Latin) Graceful. #7

Graciela (Latin) Graceful. #2

Gratia (Latin) Graceful. #2

Gratiana (Latin) Graceful. #8

Grazia (Latin) Graceful. #8

Greer (Latin) Watchman. #8

Greta (Latin) A pearl. #6

Gretchen (Latin) A pearl. #8

Grete (Latin) A pearl. #1

Gretel (Latin) A pearl. #4

Gretna (Latin) A pearl. #2

Grier (Latin) Watchman. #3

Griselda (Old German) Grey battle maiden. #3

Grizelda (Old German) Grey battle maiden. #1

Guinevere (Old Welsh) White-browed. #7

Gussie (Swedish) Staff of the Goths. #8

Gwen (22/4) (Old Welsh) White-browed. #4

Gwenda (Old Welsh) White-browed. #9

Gwendoline (Old Welsh) White-browed. #9

Gwendolyn (Old Welsh) White-browed. #2

Gwyn (Old Welsh) White-browed. #6

Gwyneth (Old Welsh) White-browed. #3

Gwynne (Old Welsh) White-browed. #7

Gypsy (Old English) Wanderer. #2

Hadara (Hebrew) Adorned with beauty. #6

Hadassah (Hebrew) Myrtle tree. #7

Hadrian (Latin) From Adria. #1

Haidee (Greek) Well-behaved. #5

Halia (22/4) (Hawaiian) Remembrance of a beloved. #4

Halima (Arabic) Gentle. #8

Hana (Hebrew) Full of grace. #6

Hanako (Japanese) Flower child. #5

Hanna (Hebrew) Full of grace. #2

Hannah (Hebrew) Full of grace. #1

Harlene (Old English) From the long field. #9

Harmonia (Latin) Harmony. #7

Harmonie (Latin) Harmony. #2

Harmony (Latin) Harmony. #4

Harriet (French) Mistress of the household. #7

Hasina (Swahili) Good. #7

Hattie (French) Mistress of the household. #9

Hatty (French) Mistress of the household. #2

Hazel (English) Hazelnut tree. #7

Heather (Old English) Flowering heather. #2

Hedda (22/4) (Old German) Strife. #4

Hedwig (Old German) Strife. #2

Hedwiga (Old German) Strife. #3

Heidi (Old German) Of noble rank. #8

Helen (Greek) Light, a torch. #8

Helena (Greek) Light, a torch. #9

Helga (Norse) Holy. #6

Helise (Old German) Famous woman warrior. #4

Helma (Old German) Helmet. #3

Heloisa (33/6) (Old German) Famous woman warrior. #6

Heloise (Old German) Famous woman warrior. #1

Henrietta (French) Mistress of the household. #1

Hermione (Greek) Of the earth. #6

Hertha (33/6) (Old English) Of the earth. #6

Hesper (Greek) Evening star. #8

Hester (Persian) A star. #3

Hiberna (Latin) Girl from Ireland. #3

Hilda (Old German) Battle maid. #7

Hilde (Old German) Battle maid. #2

Hildegarde (Old German) Battle maid. #1

Hildy (Old German) Battle maid. #4

Hinda (Hebrew) Female deer. #9

Honey (Old English) Sweet. #4

Honor (Latin) Honorable. #7

Honora (Latin) Honorable. #8

Honore (Latin) Honorable. #3

Honoria (Latin) Honorable. #8

Hope (Old English) Cheerful optimism. #8

Horatia (Latin) Keeper of the hours. #9

Hortense (Latin) Gardener. #5

Hortensia (Latin) Gardener. #1

Hosanna (Greek) A prayer of acclamation and praise. #9

Huette (Old English) Brilliant thinker. #7

Hulda (Austrian) Gracious. #1

Hyacinth (Greek) Hyacinth flower. #7

Hylda (Old German) Battle maid. #5

Hynda (Hebrew) Female deer. #7

Ianthe (Greek) Violet-colored flower. #3

Ida (Old German) Industrious. #5

Idalia (Greek) Behold the sun. #9

Idelisa (Celtic) Bountiful. #5

Idella (Celtic) Bountiful. #7

Idelle (Celtic) Bountiful. #2

Iduna (22/4) (Old Norse) Lover. #4

Ignacia (Latin) Fiery or ardent. #8

Ilana (Hebrew) Tree. #1

Ileana (Greek) Light, a torch. #6

Ilene (Greek) Light, a torch. #9

Iliana (Greek) Light, a torch. #1

Ilima (Hawaiian) The flower of Oahu. #8

Illana (Hebrew) Tree. #4

Illiana (Greek) Light, a torch. #4

Ilona (Greek) Light, a torch. #6

Ilsa (Hebrew) God's promise. #5

Ilse (Hebrew) God's promise. #9

Imelda (Old German) Powerful fighter. #8

Imena (African) A dream. #6

Imogen (Latin) Image. #9

Imogene (Latin) Image. #5

Ina (Greek) Pure. #6

India (English) From India. #1

Indira (Hindi) God of power. #1

Indra (Hindi) God of power. #1

Indria (Hindi) God of power. #1

Inez (Greek) Pure, chaste. #9

Inga (22/4) (Norse) Hero's daughter. #4

Inge (Norse) Hero's daughter. #8

Ingeborg (Norse) Hero's daughter. #5

Inger (Norse) Hero's daughter. #8

Ingrid (Norse) Hero's daughter. #7

Iola (Greek) Violet-colored dawn. #1

Iolana (Greek) Violet-colored dawn. #7

Iona (Greek) Violet-colored stone. #3

Ione (Greek) Violet-colored stone. #7

Iphigenia (Greek) Sacrifice. #6

Irena (Greek) Goddess of peace. #2

Irene (33/6) (Greek) Goddess of peace. #6

Iris (Greek) Rainbow. #1

Irma (Latin) High-ranking person. #5

Isabeau (22/4) (Old Spanish) Consecrated to God. #4

Isabel (Old Spanish) Consecrated to God. #3

Isabella (Old Spanish) Consecrated to God. #7

Isabelle (Old Spanish) Consecrated to God. #2

Isadora (Greek) Gift of Isis. #4

Isha (Hebrew) Woman. #1

Ishmaela (Hebrew) God will hear. #5

Isis (Egyptian) Supreme goddess. #2

Isleen (Gaelic) Dream or vision. #1

Isola (Latin) Isolated, a loner. #2

Isolda (Welsh) Fair lady. #6

Isolde (Welsh) Fair lady. #1

Italia (Latin) From Italy. #7

Iva (Old French) Yew tree. #5

Ivana (Hebrew) God's gracious gift. #2

Ivette (Old French) Archer with a yew bow. #9

Ivey (Old English) Ivy vine. #7

Ivonne (Old French) Archer with a yew bow. #7

Ivory (Latin) Made of ivory. #8

Ivy (Old English) Ivy vine. #2

Jacinda (Greek) Hyacinth flower. #6

Jacinta (22/4) (Greek) Hyacinth flower. #4

Jacinth (Greek) Hyacinth flower. #2

Jacoba (Hebrew) The supplanter. #5

Jacqueline (Hebrew) The supplanter. #7

Jacquelyn (Hebrew) The supplanter. #9

Jacquetta (Hebrew) The supplanter. #8

Jade (11/2) (Spanish) Jade stone. #2

Jameelah (Arabic) Beautiful. #1

Jamila (Arabic) Beautiful. #1

Jamilla (22/4) (Arabic) Beautiful. #4

Jamima (Hebrew) A dove. #2

Jana (Hebrew) The supplanter. #8

Jane (Hebrew) God's gift of grace. #3

Janeen (22/4) (Hebrew) God's gift of grace. #4

Janet (Hebrew) God's gift of grace. #5

Janette (Hebrew) God's gift of grace. #3

Janice (Hebrew) God's gift of grace. #6

Janine (Hebrew) God's gift of grace. #8

Janis (Hebrew) God's gift of grace. #8

Janna (Arabic) A harvest of fruit. #4

Janthina (Greek) Violet-colored flower. #5

Jardena (Hebrew) Descending. #8

Jarietta (Arabic) Earthen water jug. #3

Jarita (Arabic) Earthen water jug. #5

Jasmin (Arabic) Fragrant flower. #3

Jasmine (Arabic) Fragrant flower. #8

Javiera (Arabic) Bright. #3

Jayne (Hebrew) God's gift of grace. #1

Jeanelle (Hebrew) God's gift of grace. #1

Jeanette (Hebrew) God's gift of grace. #8

Jeanne (22/4) (Hebrew) God's gift of grace. #4

Jeannette (Hebrew) God's gift of grace. #4

Jeannine (Hebrew) God's gift of grace. #9

Jelena (Greek) Light, a torch. #2

Jemima (Hebrew) A dove. #6

Jemma (Latin) Precious stone. #6

Jena (Arabic) A small bird. #3

Jenevieve (French) Pure white wave. #7

Jenifer (French) Pure white wave. #4

Jenna (Arabic) A small bird. #8

Jennie (Hebrew) God's gift of grace. #3

Jennifer (French) Pure white wave. #9

Jenny (Hebrew) God's gift of grace. #5

Jeovana (Hebrew) God's gracious gift. #5

Jeraldine (Old German) Spear-mighty. #6

Jeri (Latin) Sacred or holy name. #6

Jerusha (Hebrew) The married one. #1

Jesenia (Arabic) Flower. #9

Jessamine (Arabic) Fragrant flower. #5

Jessenia (Arabic) Flower. #1

Jessica (Hebrew) Wealthy. #3

Jessika (Hebrew) Wealthy. #2

Jetta (11/2) (English) Intensely black. #2

Jewel (Old French) A precious gem. #1

Jezebel (Hebrew) Unexalted, impure. #2

Jill (Greek) Young in heart and mind. #7

Jillian (Greek) Young in heart and mind. #4

Jilly (Greek) Young in heart and mind. #5

Jina (Latin) Queen. #7

Jinny (Latin) Virginal. #9

Jinx (Latin) Charming spell. #3

Jiselle (Old German) Pledge or hostage. #9

Joann (Hebrew) God's gift of grace. #9

Joanna (Hebrew) God's gift of grace. #1

Joanne (Hebrew) God's gift of grace. #5

Joaquina (Hebrew) Jehovah has established. #7

Joby (Hebrew) The afflicted. #7

Joceline (Latin) Fair and just. #1

Jocelyn (Latin) Fair and just. #3

Jocelyne (Latin) Fair and just. #8

Joelle (Hebrew) Jehovah is God. #5

Johanna (Hebrew) God's gift of grace. #9

Johnna (Hebrew) God's gracious gift. #8

Joleen (Hebrew) He shall add. #7

Jolene (Hebrew) He shall add. #7

Joletta (Greek) Young in heart and mind. #2

Jolie (French) Pretty. #6

Jonina (Israeli) Little dove. #9

Jordana (Hebrew) Down-flowing. #9

Jorgina (Greek) Landholder, farmer. #2

Josceline (Latin) Fair and just. #2

Joselyn (Latin) Fair and just. #1

Josephine (Hebrew) He shall add. #2

Josette (22/4) (Hebrew) He shall add. #4

Josie (22/4) (Hebrew) He shall add. #4

Josilyn (Latin) Fair and just. #5

Jovita (Latin) Father of the sky. #5

Joy (Latin) Rejoicing. #5

Joyce (22/4) (Latin) Joyous. #4

Joye (Latin) Rejoicing. #1

Juana (11/2) (Hebrew) God's gracious gift. #2

Juanita (22/4) (Hebrew) God's gracious gift. #4

Judi (Hebrew) Admired. #8

Judith (Hebrew) Admired. #9

Judy (Hebrew) Admired. #6

Julia (Greek) Young in heart and mind. #8

Juliana (Greek) Young in heart and mind. #5

Juliet (Greek) Young in heart and mind. #5

June (Latin) Young. #5

Juno (Latin) Wife of Jupiter. #6

Justina (22/4) (Latin) Just. #4
Justine (Latin) Just. #8
Kachine (33/6) (American Indian) Sacred dancer. #6
Kady (Greek) Pastoral simplicity and happiness. #5
Kaila (Israeli) The laurel crown. #7
Kaitlin (Greek) Pure maiden. #4
Kaitlyn (Greek) Pure maiden. #2
Kala (Hindi) Black, time. #7
Kalani (Hawaiian) The sky. #3
Kali (Sanskrit) Black goddess, energy. #6
Kalie (Greek) Beautiful. #2
Kalifa (22/4) (Somali) Chaste, holy. #4
Kalil (Arabic) Beloved. #9
Kalila (Arabic) Beloved. #1
Kaliliah (Arabic) Beloved. #9
Kalinda (Sanskrit) Sun. #7
Kallan (Scandinavian) Flowing water. #6
Kallie (Scandinavian) Flowing water. #5
Kallista (22/4) (Greek) Most beautiful woman. #4
Kama (Sanskrit) Love. #8
Kamaria (African) Moonlike. #9
Kameko (Japanese) Child of the tortoise. #2
Kami (Japanese) Lord. #7

Kamilah (Arabic) Perfect. #1
Kanani (Hawaiian) The beautiful one. #5
Kandace (Greek) Glittering, glowing white. #3
Kani (Hawaiian) Sound. #8
Kanya (Hindi) Virgin. #7
Kara (Greek) Pure maiden. #4
Karen (22/4) (Greek) Pure maiden. #4
Kari (Greek) Pure maiden. #3
Karida (Arabic) Virgin. #8
Karima (Arabic) Noble, exalted. #8
Karin (Greek) Pure maiden. #8
Karina (Greek) Pure maiden. #9
Karis (22/4) (Greek) Grace. #4
Karissa (Greek) Very dear. #6
Karla (Old German) Little woman born to command. #7
Karma (Sanskrit) Actions are fate. #8
Karmel (Hebrew) God's vineyard. #6
Karmen (Latin) Song. #8
Karmina (Latin) Song. #4
Karyn (Greek) Pure maiden. #6
Kassandra (Greek) Disbelieved by men. #7
Kassia (Greek) Sweet-scented spice. #6
Kassie (Greek) Disbelieved by men. #1

Kate (Greek) Pure maiden. #1

Katelyn (Greek) Pure maiden. #7

Katha (Greek) Pure maiden. #5

Katharine (Greek) Pure maiden. #6

Katherine (Greek) Pure maiden. #1

Kathleen (Greek) Pure maiden. #4

Kathryn (Greek) Pure maiden. #7

Kathy (Greek) Pure maiden. #2

Katie (Greek) Pure maiden. #1

Katleen (Greek) Pure maiden. #5

Katrina (Greek) Pure maiden. #2

Katrine (33/6) (Greek) Pure maiden. #6

Kay (Greek) Pure maiden. #1

Kaya (11/2) (Japanese) Adds a place of resting. #2

Kayla (Greek) Pure maiden. #5

Kaylee (Greek) Pure maiden. #5

Keala (Hawaiian) The pathway. #3

Keeley (Gaelic) Beautiful. #9

Keely (22/4) (Gaelic) Beautiful. #4

Keiki (Hawaiian) Child. #9

Keiko (Japanese) Adored. #6

Keilani (Hawaiian) Glorious chief. #7

Keitha (Celtic) From the forest. #9

Kelila (Hebrew) Crown, laurel. #5

Kelsie (Old Norse) From the ship island. #7

Kelula (Hebrew) Crown, laurel. #8

Kendra (Old English) Knowledgeable. #8

Keren (Greek) Pure maiden. #8

Kerrin (Greek) Pure maiden. #3

Kesia (African) Favorite. #9

Ketifa (Arabic) Flowering. #7

Keturah (Hebrew) Sacrifice. #3

Kevina (Gaelic) Gentle, kind, and loving. #8

Keziah (33/6) (Greek) Sweet-scented spice. #6

Kiah (African) Season's beginning. #2

Kiara (22/4) (Greek) Pure maiden. #4

Kima (Japanese) Happiness. #7

Kimberley (Old English) From the king's wood. #1

Kimi (Japanese) Happiness. #6

Kimiko (Japanese) Heavenly child. #5

Kina (Hawaiian) China. #8

Kioko (Japanese) Child born with happiness. #7

Kiona (American Indian) Brown hills. #5

Kira (Old Persian) The sun. #3
Kirima (Eskimo) A hill. #7
Kirsten (33/6) (Norse) Anointed. #6
Kirsty (Norse) Anointed. #3
Kishi (Japanese) Happiness to the earth. #2
Kitra (Hebrew) Crowned. #5
Kittie (Greek) Pure maiden. #2
Kitty (22/4) (Greek) Pure maiden. #4
Klara (Latin) Bright, shining girl. #7
Klarissa (Latin) Bright, shining girl. #9
Koemi (Japanese) A little smile. #8
Kora (Greek) The maiden. #9
Korah (Greek) The maiden. #8
Korin (Greek) The maiden. #4
Kourtney (Old English) From the court. #3
Kristen (33/6) (Greek) Anointed. #6
Kristian (Greek) Anointed. #2
Kristie (Greek) Anointed. #1
Kristina (Greek) Anointed. #2
Kristine (Greek) Anointed. #6
Kristyn (Greek) Anointed. #8
Krystal (Latin) Without deception. #7
Kumiko (Japanese) Companion child. #8
Kyla (Gaelic) Handsome. #4

Kyna (Gaelic) Great wisdom. #6
Kyoko (Japanese) Mirror. #5
Kyra (Greek) Enthroned. #1
Lace (English) Ornamental trimming. #3
Lacey (Greek) Cheerful. #1
Ladonna (French) The lady. #7
Laila (Arabic) Dark as night. #8
Laina (Old English) From the narrow road. #1
Laine (Old English) From the narrow road. #5
Lainey (Greek) light, a torch. #3
Lala (Slavic) Tulip flower. #8
Lalita (Sanskrit) Without guile. #1
Lana (Gaelic) Comely, cheerful. #1
Lani (Hawaiian) Sky. #9
Lara (Latin) Shining, famous. #5
Laraine (33/6) (Latin) Crowned with laurels. #6
Larissa (Greek) Cheerful. #7
Lark (English) Skylark. #6
Latasha (Latin) Natal day. #8
Latisha (Latin) Gladness. #7
Latoya (Latin) Victorious. #2
Latrice (Latin) Happiness. #5
Latricia (Latin) Happiness. #1
Laura (Latin) Crowned with laurels. #8

Laurel (Latin) Laurel tree. #6

Lavelle (Latin) Cleansing. #6

Laverne (Latin) Springlike. #5

Lavina (Latin) Purified. #5

Lavinia (Latin) Purified. #5

Layla (Arabic) Dark as night. #6

Laylah (Arabic) Dark as night. #5

Lea (Hebrew) Weary. #9

Leah (Hebrew) Weary. #8

Leala (French) Loyal, faithful. #4

Leandra (Greek) Like a lioness. #1

Leane (Old French) The vine. #1

Leanna (Gaelic) Loving. #2

Leatrice (Hebrew-Latin) Weary bringer of joy. #1

Leda (Latin) Gladness. #4

Leeza (22/4) (Hebrew) God's promise. #4

Leila (Arabic) Dark as night. #3

Leilani (Hawaiian) Heavenly flower. #8

Lemuela (Hebrew) Dedicated to God. #6

Lena (Greek) Light. #5

Lenora (Greek) Light. #2

Leona (Latin) Lion. #2

Leonie (33/6) (Latin) Lion. #6

Leonora (Greek) Light. #8

Leonore (Greek) Light. #3

Leontine (Latin) Lionlike. #4

Leontyne (Latin) Lionlike. #2

Leora (Old French) Light. #6

Leorah (Old French) Light. #5

Leotie (American Indian) Prairie flower. #3

Lesenia (Arabic) Flower. #2

Leta (11/2) (Latin) Gladness. #2

Leticia (Latin) Gladness. #5

Letitia (Latin) Gladness. #4

Lettice (Latin) Gladness. #2

Levana (Hebrew) Moon or white. #1

Lexie (Greek) Helper of humankind. #1

Lexine (33/6) (Greek) Helper of humankind. #6

Leyla (Arabic) Dark as night. #1

Lia (Hebrew) Weary. #4

Lian (Chinese) The graceful willow. #9

Liana (French) A climbing vine. #1

Lianna (French) A climbing vine. #6

Libby (Hebrew) Consecrated of God. #5

Liberty (English) Freedom. #1

Lida (Greek) A woman from Lydia. #8

Liesel (Hebrew) God's promise. #8

Liezel (33/6) (Hebrew) God's promise. #6
Lila (Latin) Lily flower. #7
Lilac (Persian) Lilac flower, bluish-purple. #1
Lili (Latin) Lily flower. #6
Lilian (Latin) Lily flower. #3
Liliane (Latin) Lily flower. #8
Lilianna (Latin) lily flower. #9
Lilias (Latin) Lily flower. #8
Lilibet (33/6) (Hebrew) God's promise. #6
Lilibeth (Latin-Hebrew) Lily flower, house of God. #5
Lilith (Arabic) Of the night. #7
Lille (Latin) Lily flower. #5
Lilli (Latin) Lily flower. #9
Lillian (33/6) (Latin) Lily flower. #6
Lilliane (Latin) Lily flower. #2
Lillie (Latin) Lily flower. #5
Lily (22/4) (Latin) Lily flower. #4
Lina (Greek) Light, a torch. #9
Linda (22/4) (Spanish) Pretty. #4
Lindy (Spanish) Pretty. #1
Linette (Old French) Songbird. #4
Linnea (Scandinavian) Lime tree. #1
Linnet (Old French) Songbird. #2
Linnette (Old French) Songbird. #9

Liora (Old French) Light. #1
Lisa (Hebrew) God's promise. #5
Lisabeth (Hebrew) God's promise. #4
Lisandra (33/6) (Greek) The liberator. #6
Lise (Hebrew) God's promise. #9
Lisette (Hebrew) God's promise. #9
Lisha (22/4) (Arabic) The darkness before midnight. #4
Liv (Latin) Symbol of peace. #7
Livia (Latin) Symbol of peace. #8
Liz (Hebrew) God's promise. #2
Liza (Hebrew) God's promise. #3
Lizette (Hebrew) God's promise. #7
Lois (Old German) Famous warrior maid. #1
Lola (Spanish) Sorrows. #4
Lolita (Spanish) Sorrows. #6
Lomasi (American Indian) Pretty flower. #6
Lona (Old English) Solitary. #6
Lonna (Old English) Solitary. #2
Lora (Latin) Crowned with laurels. #1
Lorelei (German) Siren. #4

Lorelle (Latin) Little. #7

Lorena (Latin) Crowned with laurels. #2

Lorenza (Latin) Crowned with laurels. #1

Loretta (Latin) Crowned with laurels. #1

Lori (Latin) Crowned with laurels. #9

Lorna (Latin) Crowned with laurels. #6

Lorraine (Latin) Crowned with laurels. #2

Lorrie (Latin) Crowned with laurels. #5

Lottie (French) Little and womanly. #9

Lotty (French) Little and womanly. #2

Lotus (Greek) Lotus flower. #6

Louella (Old English) Elf. #6

Louellen (33/6) (Old English) Elf. #6

Louisa (Old German) Famous battle maid. #5

Louise (Old German) Famous battle maid. #9

Lourdes (Basque) Rocky point. #4

Luana (Old German-Hebrew) Graceful battle maid. #4

Lucia (Latin) Light. #1

Lucianne (Latin) Crowned with laurels. #7

Lucilla (Latin) Light. #7

Lucille (Latin) Light. #2

Lucinda (Latin) Light. #1

Lucine (Latin) Light. #1

Lucrecia (Latin) Riches, reward. #9

Lucretia (Latin) Riches, reward. #8

Lucy (Latin) Light. #7

Ludella (22/4) (Old English) Pixie maid. #4

Ludmilla (Slavic) Loved by the people. #3

Luella (Old English) Elf. #9

Luellen (Old English) Elf. #9

Luisa (Old German) Famous battle maid. #8

Lulu (Old German) Famous battle maid. #3

Lumina (Latin) Brilliant, illuminated. #7

Luna (Latin) The moon. #3

Lunetta (Latin) Little moon. #3

Lupita (Arabic) River of black stones. #7

Lura (German) Siren. #7

Lurette (German) Siren. #2

Lurleen (33/6) (German) Siren. #6

Lurlene (33/6) (German) Siren. #6

Lurline (German) Siren. #1

Luz (Spanish) Light. #5

Lydia (Greek) A woman from Lydia. #6

Lynda (Spanish) Pretty. #2

Lynette (Old French) Linnet bird. #2

Lynne (Old English) From the waterfalls. #7

Lynnette (Old French) Linnet bird. #7

Lyric (Latin) Of the lyre. #4

Lysandra (Greek) The liberator. #4

Lysette (Hebrew) God's promise. #7

Mabel (Latin) Lovable. #6

Machiko (33/6) (Japanese) Beautiful child. #6

Madelaine (Hebrew) Woman of Magdala. #1

Madeleine (Hebrew) Woman of Magdala. #5

Madelena (Hebrew) Woman of Magdala. #1

Madeline (Hebrew) Woman of Magdala. #9

Madelon (Hebrew) Woman of Magdala. #1

Madge (Latin) A pearl. #3

Madonna (Italian) My lady. #8

Mae (Latin) Great. #1

Maeko (Japanese) Truth child. #9

Maemi (Japanese) Smile of truth. #5

Maeve (Celtic) Intoxicating. #1

Magali (Hebrew) Woman of Magdala. #7

Magda (Hebrew) Woman of Magdala. #8

Magdala (Hebrew) Woman of Magdala. #3

Magdalane (Hebrew) Woman of Magdala. #4

Magdalen (Hebrew) Woman of Magdala. #3

Magdalena (Hebrew) Woman of Magdala. #4

Magdalene (Hebrew) Woman of Magdala. #8

Magdalia (Hebrew) Woman of Magdala. #3

Maggee (Latin) A pearl. #2

Maggie (33/6) (Latin) A pearl. #6

Magnolia (French) Flower named after Pierre Magnol. #9

Mahala (Hebrew) Affection. #9

Mahalia (Hebrew) Affection. #9

Mahina (Hawaiian) Moon, moonlight. #1

Mai (Latin) Great. #5

Maia (Latin) Great. #6

Maida (Old English) Maiden. #1

Maire (Hebrew) Bitterness. #1

Maisey (Latin) A pearl. #9

Maisie (Latin) A Pearl. #2

Majesta (Latin) Royal bearing. #6

Malana (Hawaiian) Buoyant, light. #6

Malia (Hebrew) Bitterness. #9

Malka (11/2) (Hebrew) Queen. #2

Malva (Greek) Soft and tender. #4

Malvina (Greek) Soft and tender. #9

Mame (Hebrew) Bitterness. #5

Mamie (Hebrew) Bitterness. #5

Mandi (Latin) Worthy of being loved. #5

Mandie (Latin) Worthy of being loved. #1

Mandisa (African) Sweetness. #7

Mandy (Latin) Worthy of being loved. #3

Manuela (22/4) (Hebrew) God is with us. #4

Mara (Latin) Of the sea. #6

Marcella (Latin) Follower of Mars. #2

Marcia (Latin) Follower of Mars. #9

Marcie (Latin) Follower of Mars. #4

Marcy (Latin) Follower of Mars. #6

Mare (Hebrew) Bitterness. #1

Margaret (Latin) A pearl. #2

Margareta (Latin) A pearl. #3

Margaretta (Latin) A pearl. #5

Margarita (Latin) A pearl. #7

Marge (Latin) A pearl. #8

Margery (Latin) A pearl. #6

Margo (Latin) A pearl. #9

Margot (Latin) A pearl. #2

Marguerita (Latin) A pearl. #5

Marguerite (Latin) A pearl. #9

Mari (Hebrew) Bitterness. #5

Maria (Hebrew) Bitterness. #6

Mariah (Hebrew) Bitterness. #5

Marian (Hebrew) Bitter, graceful. #2

Mariana (Hebrew) Bitter, graceful. #3

Mariann (Hebrew) Bitter, graceful. #7

Marianne (Hebrew) Bitter, graceful. #3

Maribel (33/6) (Hebrew-French) Bitter, beautiful. #6

Maribelle (Hebrew-French) Bitter, beautiful. #5

Maricela (Latin) Of the sea. #8

Marie (Hebrew) Bitterness. #1

Mariel (Hebrew) Bitterness. #4

Mariela (Hebrew) Bitterness. #5

Marietta (33/6) (Hebrew) Bitterness. #6

Marigold (English) Mary's gold. #7

Mariko (Japanese) Ball, circle. #4

Marilla (Latin) Shining sea. #3

Marilyn (Hebrew) Bitterness. #2

Marin (Latin) From the sea. #1

Marina (Latin) From the sea. #2

Maris (Latin) Of the sea. #6

Marisa (Latin) Of the sea. #7

Marise (Japanese) Infinite, endless. #2

Marisel (Latin) Of the sea. #5

Marisela (33/6) (Latin) Of the sea. #6

Marisol (33/6) (Hebrew-Latin) Bitterness, alone. #6

Marissa (Latin) Of the sea. #8

Marjorie (Latin) A pearl. #8

Marla (Hebrew) Bitterness. #9

Marlene (Hebrew) Bitterness. #5

Marney (Latin) From the sea. #4

Marnie (33/6) (Latin) From the sea. #6

Marsha (Latin) Follower of Mars. #6

Martha (Arabic) Lady, mistress. #7

Martina (Latin) Follower of Mars. #4

Martita (Arabic) Lady, mistress. #1

Marvella (Latin) Wonderful, extraordinary. #3

Marvina (33/6) (Old English) Lover of the sea. #6

Mary (Hebrew) Bitterness. #3

Maryann (Hebrew) Bitter, graceful. #5

Maryanne (Hebrew) Bitter, graceful. #1

Maryleen (Hebrew) Bitterness. #3

Marylin (Hebrew) Bitterness. #2

Marylou (33/6) (Hebrew-Old German) Bitter battle maid. #6

Mata (Hebrew) Gift of God. #8

Mathilda (Old German) Powerful in battle. #5

Matilda (Old German) Powerful in battle. #6

Matsuko (Japanese) Pine tree child. #1

Maud (Old German) Powerful in battle. #3

Maude (Old German) Powerful in battle. #8

Maura (Hebrew) Bitterness. #9

Maureen (Hebrew) Bitterness. #5

Maurene (Hebrew) Bitterness. #5

Mauve (French) A purple-colored mallow plant. #8

Mavis (French) Thrush. #1

Maximilia (Latin) Most excellent. #1

Maxine (Latin) Most excellent. #3

May (Hebrew) Bitterness. #3

Maya (Sanskrit) Illusion. #4

Meagan (Gaelic) Soft, gentle. #5

Meagen (Gaelic) Soft, gentle. #9

Media (Greek) Ruling. #5

Meg (Latin) A woman without equal. #7

Megan (22/4) (Gaelic) Soft, gentle. #4

Meghan (Gaelic) Soft, gentle. #3

Meiko (Japanese) A bud. #8

Melanie (Greek) Dark-clothed. #5

Melantha (Greek) Lady of the night. #2

Melba (Latin) Mallow flower. #6

Melina (Latin) Canary-yellow-colored. #9

Melinda (Greek) Dark-clothed. #4

Melisande (Greek) Honeybee. #1

Melissa (Greek) Honeybee. #6

Melita (Hebrew) God's vineyard. #6

Melly (22/4) (Greek) Dark-clothed. #4

Melodie (Greek) Song. #9

Melody (Greek) Song. #2

Melora (Latin) Make better. #1

Melvina (Gaelic) Brilliant chief. #4

Mercedes (Spanish) Compassion. #9

Mercia (Latin) Compassion. #4

Mercy (Latin) Compassion. #1

Merna (Gaelic) Beloved. #6

Merrie (English) Mirthful, joyous. #5

Merry (English) Mirthful, joyous. #7

Mertice (Old English) Famous and pleasant. #1

Meryl (French) Blackbird. #1

Messina (Latin) The middle child. #8

Mia (Latin) Mine. #5

Michaela (Hebrew) Like unto the Lord. #7

Micheline (Hebrew) Like unto the Lord. #6

Michelle (Hebrew) Like unto the Lord. #4

Michie (Japanese) Gracefully drooping flower. #2

Michiko (Japanese) Beauty and wisdom. #5

Midori (Japanese) Green. #5

Miki (Japanese) Three trees together. #6

Milagros (Latin) Miracles. #4

Mildred (Old English) Gentle strength. #2

Miliani (Hawaiian) Gentle caress. #4

Millicent (Old German) Industrious. #7

Millie (33/6) (Old German) Industrious. #6

Milly (Old English) Gentle strength. #8

Mimi (Hebrew) Bitterness. #8

Mina (Old German) Love, tender affection. #1

Mindy (Greek) Dark, gentle. #2

Minerva (Greek) Wisdom. #1

Minette (Old German) Love, tender affection. #5

Minna (Old German) Love, tender affection. #6

Minnie (Old German) Love, tender affection. #1

Mira (Latin) Wonderful. #5

Mirabel (33/6) (Latin) Of extraordinary beauty. #6

Mirabella (Latin) Of extraordinary beauty. #1

Miranda (33/6) (Latin) Admirable. #6

Mirella (Latin) Wonderful. #7

Mireya (Latin) Wonderful. #8

Miriam (Hebrew) Bitterness. #9

Missy (22/4) (Greek) Honeybee. #4

Misty (Old English) Covered with mist. #5

Mitzi (Hebrew) Bitterness. #5

Miya (Japanese) Three arrows. #3

Modesty (Latin) Modest, moderate. #2

Mohala (Hawaiian) Petals unfolding, shining forth. #9

Moira (Hebrew) Bitterness. #2

Mollie (Hebrew) Bitterness. #3

Molly (Hebrew) Bitterness. #5

Mona (Greek) Solitary. #7

Monica (Latin) Advisor. #1

Monique (Latin) Advisor. #4

Monserrat (Latin) Jagged mountain. #6

Morag (Celtic) Great. #9

Morena (Spanish) Brown, brown-haired. #3

Morgaine (Scottish Gaelic) From the edge of the sea. #1

Morgana (33/6) (Scottish Gaelic) From the edge of the sea. #6

Moriah (Hebrew) God is my teacher. #1

Morisa (Latin) Dark-skinned. #3

Morwenna (Welsh) Ocean waves. #4

Moselle (Hebrew) Drawn out of the water. #9

Moyra (Celtic) Great. #9

Mozelle (Hebrew) Drawn out of the water. #7

Muriel (33/6) (Hebrew) Bitterness. #6

Musetta (French) Child of the Muses. #9

Musidora (Greek) Gift of the Muses. #1

Myisha (Arabic) Woman, life. #3

Myra (Greek) Fragrant ointment. #3

Myrna (Gaelic) Polite, gentle. #8

Myrta (Greek) Myrtle plant. #5

Myrtle (Greek) Myrtle plant. #3

Mystica (French) Air of mystery. #9

Nada (11/2) (Russian) Hope. #2

Nadeen (Russian) Hope. #7

Nadia (Russian) Hope. #2

Nadine (Russian) Hope. #2

Nadya (Russian) Hope. #9

Nalani (Hawaiian) Calmness of the heavens. #6

Nan (11/2) (Hebrew) Full of grace. #2

Nana (Hebrew) Full of grace. #3

Nancee (Hebrew) Full of grace. #6

Nanci (Hebrew) Full of grace. #5

Nancie (Hebrew) Full of grace. #1

Nancy (Hebrew) Full of grace. #3

Nanette (Hebrew) Full of grace. #7

Nani (Hawaiian) Beautiful. #2

Nanice (Hebrew) Full of grace. #1

Nanine (Hebrew) Full of grace. #3

Nannette (Hebrew) Full of grace. #3

Nannie (Hebrew) Full of grace. #3

Nanny (Hebrew) Full of grace. #5

Naomi (Hebrew) Pleasant. #7

Nara (American Indian) Oak. #7

Nari (Japanese) Thunderpeal. #6

Nariko (Japanese) Thunderpeal. #5

Nasaya (Hebrew) Miracle of God. #7

Nastassia (22/4) (Latin) Natal day. #4

Natala (Latin) Natal day. #4

Natalee (22/4) (Latin) Natal day. #4

Natalia (22/4) (Latin) Natal day. #4

Natalie (Latin) Natal day. #8

Natalya (Latin) Natal day. #2

Natasha (Latin) Natal day. #1

Nathalia (Latin) Natal day. #3

Nathalie (Latin) Natal day. #7

Nathania (Hebrew) Gift of God. #5

Natividad (Latin) Natal day. #3

Nazneen (Persian) Exquisitely beautiful, charming. #7

Neala (Gaelic) The champion. #6

Neda (Old English) Rich and happy protector. #6

Nedda (Old English) Rich and happy protector. #1

Neema (Swahili) Born in prosperity. #2

Neila (Gaelic) The champion. #5

Neile (Gaelic) The champion. #9

Neilla (Gaelic) The champion. #8

Neille (Gaelic) The champion. #3

Neiva (Spanish) Snow. #6

Nelda (Old English) Born under an elder tree. #9

Nelia (Latin) Yellow, horn-colored. #5

Nell (Old French) Light. #7

Nella (Old French) Light. #8

Nellie (Old French) Light. #3

Nellwyn (33/6) (Greek) Friend and companion. #6

Nelly (Old French) Light. #5

Neoma (Greek) New moon. #3

Neona (22/4) (Greek) New moon. #4

Nereida (Greek) Sea nymph. #2

Neri (Greek) Of the sea. #1

Nerima (33/6) (Greek) Of the sea. #6

Nerine (Greek) Sea nymph. #2

Nerissa (Greek) Of the sea. #4

Nerita (Greek) Of the sea. #4

Nesta (Greek) Pure. #5

Netta (Hebrew) Full of grace. #6

Nettie (Hebrew) Full of grace. #1

Netty (Hebrew) Full of grace. #3

Neysa (Greek) Pure. #1

Niabi (American Indian) A fawn. #8

Nichola (Greek) The people's victory. #8

Nichole (Greek) The people's victory. #3

Nicola (Greek) The people's victory. #9

Nicole (Greek) The people's victory. #4

Nicolette (Greek) The people's victory. #4

Nidia (Latin) From the nest. #1

Nina (Spanish) Girl. #2

Ninetta (Spanish) Girl. #2

Ninette (33/6) (Spanish) Girl. #6

Ninon (Spanish) Girl. #3

Nissa (Norse) Friendly elf or brownie. #8

Nisse (Norse) Friendly elf or brownie. #3

Nissie (Norse) Friendly elf or brownie. #3

Nissy (Norse) Friendly elf or brownie. #5

Nita (American Indian) Bear. #8

Noelani (Hawaiian) Beautiful one from heaven. #7

Noella (French) Born at Christmas. #5

Noelle (French) Born at Christmas. #9

Noellyn (French) Born at Christmas. #7

Noemi (Hebrew) Pleasant. #2

Nola (Latin) Symbol of peace. #6

Nolana (Latin) Symbol of peace. #3

Noleta (22/4) (Latin) Unwilling. #4

Nona (Latin) The ninth. #8

Nonie (Latin) The ninth. #3

Nora (Greek) Light. #3

Norah (Greek) Light. #2

Noreen (Latin) A rule, pattern, or precept. #8

Norell (Scandinavian) From the north. #4

Noriko (Japanese) Child of ceremony. #1

Norina (Latin) A rule, pattern, or precept. #8

Norine (Latin) A rule, pattern, or precept. #3

Norma (Latin) A rule, pattern, or precept. #7

Nova (Latin) New. #7

Nuri (Arabic) Light. #8

Nydia (Latin) From the nest. #8

Nyla (Gaelic) the champion. #7

Nyssa (Norse) Friendly elf or brownie. #6

Oceana (Greek) Ocean. #3

Octavia (Latin) The eighth. #8

Odele (Hebrew) I will praise God. #5

Odelia (Hebrew) I will praise God. #1

Odelinda (Hebrew) I will praise God. #1

Odell (Hebrew) I will praise God. #3

Odella (22/4) (Hebrew) I will praise God. #4

Odessa (Greek) Wandering, quest. #9

Odetta (Hebrew) I will praise God. #2

Odette (French) Home lover. #6

Ofelia (Greek) Serpent. #3

Ola (Hawaiian) Life, well-being.
#1

Oleen (Hawaiian) Joyous. #6

Oletha (Scandinavian) Light,
nimble. #7

Olexa (Greek) Helper of
humankind. #3

Olga (Norse) Holy. #8

Olia (Latin) Symbol of peace.
#1

Oliana (Polynesian) Oleander.
#7

Olida (Latin) Symbol of peace.
#5

Olina (Hawaiian) Joyous. #6

Olive (Latin) Symbol of peace.
#9

Olivia (Latin) Symbol of peace.
#5

Olwen (Welsh) White footprint.
#6

Olwyn (Welsh) White footprint.
#8

Olympia (Greek) Heavenly. #1

Omega (Greek) Ultimate. #5

Ona (Latin) Unity. #3

Ondine (Latin) Wave, of water.
#7

Ondrea (Latin) Womanly. #3

Oneida (American Indian) The
looked-for one. #3

Onida (American Indian) The
looked-for one. #7

Oona (Latin) Unity. #9

Opal (Sanskrit) A precious
stone. #8

Opalina (Sanskrit) A precious
stone. #5

Opaline (Sanskrit) A precious
stone. #9

Ophelia (Greek) Serpent. #3

Ophelie (Greek) Serpent. #7

Ophra (Hebrew) Young deer.
#4

Oprah (Hebrew) Young deer.
#4

Orah (Israeli) Light. #6

Oralia (Latin) Golden. #2

Oralie (33/6) (Latin) Golden.
#6

Ordelia (Old German) Elf's
spear. #1

Orelia (33/6) (Latin) Golden.
#6

Orenda (American Indian)
Magical power. #3

Oria (Israeli) Light. #7

Oriana (Latin) Dawning,
golden. #4

Orianna (Latin) Dawning,
golden. #9

Ottilie (Old German)
Prosperous. #9

Palila (Polynesian) Bird. #6

Palma (Latin) The palm-bearer.
#7

Palmira (Latin) The palm-
bearer. #7

Paloma (22/4) (Spanish) Dove. #4

Pam (Greek) All honey. #3

Pamela (Greek) All honey. #3

Pamelina (Greek) All honey. #8

Pamella (Greek) All honey. #6

Pammy (Greek) All honey. #5

Pamona (Latin) Fertile. #6

Pandora (33/6) (Greek) All-gifted. #6

Pansey (Greek) Fragrant. #8

Pansie (Greek) Fragrant. #1

Pansy (Greek) Flowerlike. #3

Paola (Latin) Little. #9

Paolina (Latin) Little. #5

Parthena (Greek) Virgin. #2

Pascale (Latin) Born at Easter or Passover. #3

Patci (22/4) (Latin) Wellborn, noble. #4

Patience (French) Endurance with fortitude. #1

Patrica (Latin) Wellborn noble. #5

Patricia (Latin) Wellborn, noble. #5

Patti (Latin) Wellborn, noble. #3

Paula (Latin) Little. #6

Paulette (Latin) Little. #1

Paulina (Latin) Little. #2

Pauline (33/6) (Latin) Little. #6

Paz (Latin) Peace. #7

Pazia (Hebrew) Golden. #8

Peace (Old English) The peaceful one. #3

Pearl (Latin) A pearl. #7

Peg (Latin) A pearl. #1

Peggy (33/6) (Latin) A pearl. #6

Pelagia (33/6) (Greek) The sea. #6

Penelope (Greek) The weaver. #7

Pennie (Greek) The weaver. #9

Penny (Greek) The weaver. #2

Peony (Greek) Praise-giving. #3

Pepita (Hebrew) He shall add. #4

Perdita (Latin) Lost. #1

Perina (Latin) Rock or stone. #9

Perla (Latin) A pearl. #7

Perrine (Latin) Rock or stone. #4

Persephone (Greek) Goddess of the underworld. #4

Peta (Latin) Rock or stone. #6

Petra (Latin) Rock or stone. #6

Petrina (Latin) Rock or stone. #2

Petronella (Latin) Rock or stone. #1

Petronia (Latin) Rock or stone. #8

Petula (Latin) Seeker. #3

Petunia (American Indian) Petunia flower. #5

Phaedra (Greek) Bright. #8

Phebe (Greek) Shining. #9

Phedra (Greek) Bright. #7

Philana (Greek) Lover of mankind. #7

Philantha (Greek) Lover of flowers. #8

Philberta (Old German) Very brilliant. #1

Philippa (Greek) Lover of horses. #6

Phillippa (Greek) Lover of horses. #9

Phillis (Greek) A green branch. #4

Philomela (Greek) Lover of song. #1

Philomena (Greek) Loving song. #3

Phoebe (33/6) (Greek) Shining. #6

Phylis (Greek) A green branch. #8

Phyllida (Greek) A green branch. #6

Phyllis (11/2) (Greek) A green branch. #2

Phyllys (Greek) A green branch. #9

Pia (Latin) Devout. #8

Pier (Latin) Rock or stone. #3

Pierette (Latin) Rock or stone. #8

Pietra (33/6) (Latin) Rock or stone. #6

Pilar (Spanish) Pillar of the church. #2

Piper (Old English) A pipe player. #1

Pippa (Greek) Lover of horses. #4

Pippy (Greek) Lover of horses. #1

Placida (Latin) Peaceful, serene. #1

Placidia (Latin) Peaceful, serene. #1

Poala (Latin) Little. #9

Polly (Latin) Little. #8

Pollyanna (Latin-Hebrew) Little, graceful. #2

Pollyanne (Latin-Hebrew) Little, graceful. #6

Pomona (Latin) Fertile. #2

Poppy (Latin) Poppy flower. #7

Portia (Latin) Offering. #7

Posy (English) Small bunch of flowers. #3

Preciosa (Old French) Of great value. #5

Precious (Old French) Of great value. #7

Prima (Latin) First or first child. #3

Primrose (Latin) First rose. #5

Princesa (Latin) Princess. #4

Princess (Latin) Princess. #4

Pris (Latin) From ancient times. #8

Prisca (Latin) From ancient times. #3

Priscilla (Latin) From ancient times. #9

Prissie (Latin) From ancient times. #5

Prospera (Latin) Favorable. #9

Prudence (Latin) Foresight. #5

Prudie (Latin) Foresight. #1

Prudy (Latin) Foresight. #3

Prue (Latin) Foresight. #6

Prunella (Latin) Plum color. #9

Psyche (Greek) Soul or mind. #4

Purity (English) Purity. #1

Queena (Old English) Queen. #9

Queenie (Old English) Queen. #4

Querida (Spanish) Beloved. #3

Questa (French) Searcher. #2

Quinta (Latin) Fifth child. #1

Quintana (Latin) Fifth child. #7

Quintessa (Latin) Fifth essence. #8

Quintilla (Latin) Fifth child. #7

Quintina (Latin) Fifth child. #6

Rabiah ((Arabic) Breeze. #3

Rachael (Hebrew) Innocent as a lamb. #3

Rachel (Hebrew) Innocent as a lamb. #2

Rachell (Hebrew) Innocent as a lamb. #5

Rachelle (Hebrew) Innocent as a lamb. #1

Radmilla (Slavic) Worker for the people. #7

Rae (Hebrew) Innocent as a lamb. #6

Raeanne (Hebrew) Innocent as a lamb. #4

Rafaela (Hebrew) Healed by God. #8

Rainbow (English) Rainbow. #1

Raine (Sanskrit) Queen. #2

Raisa (Greek) A rose. #3

Ramah (Israeli) High. #5

Ramona (Old German) Mighty or wise protector. #8

Ramonda (Old German) Mighty or wise protector. #3

Rana (Sanskrit) Queen. #7

Randa (Old English) Swift wolf. #2

Randene (Old English) Swift wolf. #7

Rani (Sanskrit) Queen. #6

Ranita (Hebrew) Song. #9

Raphaela (Hebrew) Healed by God. #8

Raquel (Hebrew) Innocent as a lamb. #2

Rasheeda (Arabic) Counselor. #7

Rashida (33/6) (Arabic) Counselor. #6

Raven (Old English) Like the raven. #6

Raya (Israeli) Friend. #9

Raylina (Hebrew) Innocent as a lamb. #8

Raymonda (Old German) Mighty or wise protector. #1

Rayna (Sanskrit) Queen. #5

Raynell (33/6) (Sanskrit) Queen. #6

Raynice (Sanskrit) Queen. #3

Rea (Greek) Flows from the earth. #6

Reanna (Welsh) A witch. #8

Reanne (Welsh) A witch. #3

Reba (Hebrew) The captivator. #8

Rebeca (Hebrew) The captivator. #7

Rebecca (Hebrew) The captivator. #1

Rebekah (Hebrew) The captivator. #5

Reena (Hebrew) Song. #7

Reeta (22/4) (Latin) A pearl. #4

Reeva (French) Riverside. #6

Regina (Latin) Queen. #9

Reiko (Japanese) Gratitude, propriety. #4

Reina (Latin) Queen. #2

Remy (French) From Rheims. #7

Rena (Hebrew) Song. #2

Renae (Latin) Reborn. #7

Renata (Latin) Reborn. #5

Renate (Latin) Reborn. #9

Renee (Latin) Reborn. #2

Renell (Latin) Reborn. #3

Renelle (Latin) Reborn. #8

Renie (33/6) (Latin) Reborn. #6

Reta (African) To shake. #8

Reva (French) Riverside. #1

Rexana (Latin) Regal and graceful. #9

Reynalda (Old English) Mighty and powerful. #8

Rhea (Greek) Flows from the earth. #5

Rheba (Hebrew) The captivator. #7

Rheta (African) To shake. #7

Rhetta (African) To shake. #9

Rhiamon (Welsh) A witch. #6

Rhianon (Welsh) A witch. #7

Rhoda (Latin) Dew of the sea. #1

Rhody (Latin) Dew of the sea. #7

Rhona (Old Norse) Rough isle. #2

Rhonda (33/6) (Welsh) Grand. #6

Riane (Welsh) A witch. #2

Riannon (Welsh) A witch. #4

Rianon (Welsh) A witch. #8

Rica (22/4) (Old German) Peaceful ruler. #4

Ricca (Old German) Peaceful ruler. #7

Rihana (33/6) (Arabic) Sweet basil. #6

Rima (Spanish) Rhyme, poetry. #5

Risa (Latin) Laughter. #2

Rita (Latin) A pearl. #3

Riva (French) Riverside. #5

Roana (22/4) (Greek-Hebrew) A rose, full of grace. #4

Roanna (Greek-Hebrew) A rose, full of grace. #9

Roanne (Greek-Hebrew) A rose, full of grace. #4

Roberta (Old English) Bright fame. #7

Robina (Old English) Bright fame. #5

Robinette (Old English) Bright fame. #9

Rochella (French) From the little rock. #2

Rochelle (French) From the little rock. #6

Rocio (33/6) (Latin) Dew. #6

Roda (Latin) Dew of the sea. #2

Roderica (Old German) Famous ruler. #1

Rodina (Old English) From the reed valley. #7

Rohana (Hindi) Sandalwood. #3

Rolanda (Old German) From the famous land. #2

Roma (Latin) Eternal city. #2

Romina (Latin) Woman of Rome. #7

Romona (Old German) Mighty or wise protector. #4

Rona (Old English) Mighty and powerful. #3

Ronalda (Old English) Mighty and powerful. #2

Ronda (Welsh) Grand. #7

Rosa (Greek) A rose. #8

Rosabel (French-Greek) Beautiful rose. #9

Rosabella (French-Greek) Beautiful rose. #4

Rosabelle (French-Greek) Beautiful rose. #8

Rosalba (Latin) White rose. #5

Rosaleen (Spanish) Beautiful rose. #8

Rosalia (Latin) Rose. #3

Rosalie (Latin) Rose. #7

Rosalind (Spanish) Beautiful rose. #2

Rosalinda (Spanish) Beautiful rose. #3

Rosalinde (Spanish) Beautiful rose. #7

Rosaline (Spanish) Beautiful rose. #3

Rosalyn (Spanish) Beautiful rose. #5

Rosalynd (Spanish) Beautiful rose. #9

Rosamond (Old German) Famous guardian. #9

Rosamund (33/6) (Old German) Famous guardian. #6

Rosanna (Greek-Hebrew) A rose, full of grace. #1

Rosanne (Greek-Hebrew) A rose, full of grace. #5

Rose (Greek) A rose. #3

Roseanne (Greek-Hebrew) A rose, full of grace. #1

Rosemaria (English) Rose of Saint Mary. #9

Rosemarie (English) Rose of Saint Mary. #4

Rosemary (English) Rose of Saint Mary. #6

Rosemund (Old German) Famous guardian. #1

Rosene (Greek) A rose. #4

Rosetta (Greek) A rose. #8

Rosette (Greek) A rose. #3

Roshelle (French) From the little rock. #4

Rosie (Greek) A rose. #3

Rosina (Greek) A rose. #4

Rosita (Greek) A rose. #1

Roslyn (Spanish) Beautiful rose. #4

Rosy (Greek) A rose. #5

Rowena (Old English) Well-known friend. #4

Roxana (Persian) Brilliant. #1

Roxanne (Persian) Brilliant. #1

Roya (Old French) Royal. #5

Royale (Old French) Royal. #4

Roza (Greek) A rose. #6

Rozina (Greek) A rose. #2

Rue (English) Regret. #8

Ruella (Hebrew-Greek) Compassionate, light. #6

Rufina (33/6) (Latin) Red-haired. #6

Rui (Japanese) Troublesome. #3

Rula (Latin) Sovereign ruler. #7

Russet (English) A reddish or yellowish-brown color. #3

Ruth (22/4) (Hebrew) Compassionate. #4

Ruthann (33/6) (Hebrew) Compassionate, full of grace. #6

Ruthanne (Hebrew) Compassionate, full of grace. #2

Ruthie (Hebrew) Compassionate. #9

Ruthy (Hebrew) Compassionate. #2

Sabina (Latin) Sabine woman. #1

Sabra (Hebrew) Thorny cactus. #5

Sabrina (Latin) A Princess. #1

Sachi (22/4) (Japanese) Bliss child, joy. #4

Sachiko (Japanese) Bliss child, joy. #3

Sade (11/2) (Hebrew) Princess. #2

Sadie (Hebrew) Princess. #2

Sadira (Persian) Lotus. #7

Sahara (Arabic) Wilderness. #3

Salena (Latin) Salty. #7

Salina (Latin) Salty. #2

Sally (Hebrew) Princess. #6

Salome (Hebrew) Peace. #2

Salvia (Latin) Sage herb. #1

Samala (11/2) (Hebrew) Asked of God. #2

Samantha (Aramaic) Listener. #5.

Samara (Hebrew) Ruled by God. #8

Samaria (Hebrew) Ruled by God. #8

Sami (Hebrew) His name is God. #6

Samuela (Hebrew) His name is God. #9

Sandra (Greek) Helper of humankind. #3

Santana (Spanish) Saint Anne. #7

Sapphira (Greek) Sapphire stone. #7

Sapphire (Greek) Sapphire stone. #2

Sara (Hebrew) Of royal status. #3

Sarah (Hebrew) Of royal status. #2

Saree (Arabic) Most noble. #3

Sari (Arabic) Most noble. #2

Sarina (Hebrew) Of royal status. #8

Sarita (Hebrew) Of royal status. #5

Sasa (Japanese) Help, aid. #4

Sascha (Greek) Helper of humankind. #6

Satin (French) Satin fabric. #9

Satina (French) Satin fabric. #1

Savannah (Spanish) Barren. #8

Savina (Latin) Sabine woman. #3

Saxona (Old German) Sword bearer. #2

Scarlet (Old English) Scarlet-colored. #6

Scarlett (Old English) Scarlet-colored. #8

Seana (Hebrew) God's gracious gift. #4

Secunda (22/4) (Latin) Second-born. #4

Seema (Hebrew) Treasure. #7

Sela (Hebrew) A rock. #1

Selena (Greek) Moon. #2

Selene (Greek) Moon. #6

Selima (Hebrew) Peaceful. #5

Selina (Greek) Moon. #6

Selma (Old Norse) Divinely protected. #5

Sephira (Hebrew) Ardent. #4

Serafina (Hebrew) Ardent. #1

Seraphina (Hebrew) Ardent. #1

Serena (Latin) Calm, serene. #8

Serenity (Latin) Calm, serene. #7

Shaina (Hebrew) Beautiful. #7

Shana (Gaelic) Small and wise. #7

Shandra (Sanskrit) Moonlike. #2

Shani (African) Marvelous. #6

Shanice (African) Marvelous. #5

Shanleigh (Gaelic) Child of the old hero. #2

Shanley (Gaelic) Child of the old hero. #3

Shantel (Latin) Song. #7

Sharde (African) Honor confers a crown. #1

Shari (Hebrew) Of royal status. #1

Sharik (African) God's child. #3

Sharleen (French) Little and womanly. #1

Sharlene (French) Little and womanly. #1

Sharmaine (Latin) Song. #7

Sharon (Hebrew) A level plain. #3

Sharona (Hebrew) A level plain. #4

Sharron (Hebrew) A level plain. #3

Sharyn (Hebrew) A level plain. #4

Shawna (Hebrew) God's gracious gift. #3

Sheba (Hebrew) From Sheba. #8

Sheela (Celtic) Musical. #5

Sheena (Hebrew) God's gift of grace. #7

Sheenah (33/6) (Hebrew) God's gift of grace. #6

Shelia (Celtic) Musical. #9

Sherilyn (Old English) From the bright meadow. #2

Sherri (Old French) Beloved. #5

Sherrie (Old French) Beloved. #1

Sherry (Old French) Beloved. #3

Sheryl (33/6) (Old English) From the bright meadow. #6

Shifra (Hebrew) Beautiful. #7

Shina (Japanese) Good, virtue. #6

Shirlee (Old English) From the bright meadow. #4

Shirleen (Old English) From the bright meadow. #9

Shirlene (Old English) From the bright meadow. #9

Shirley (Old English) From the bright meadow. #6

Shoshana (Hebrew) Graceful lily. #4

Shulamit (Hebrew) Peace. #4

Shulamith (Hebrew) Peace. #3

Sian (Hebrew) God's gracious gift. #7

Sibilla (Greek) Prophetess. #1

Sibley (Greek) Prophetess. #9

Sibyl (22/4) (Greek) Prophetess. #4

Sibylle (Greek) Prophetess. #3

Sidra (Latin) Related to the stars. #6

Sienna (Latin) A brownish-red color. #8

Sierra (Latin) Mountainous. #7

Sigrid (Old German) Victorious protector. #3

Silver (Old English) Silver colored metal. #4

Silvia (Latin) From the forest. #9

Sima (Hebrew) Treasure. #6

Simona (Hebrew) One who hears. #8

Simone (Hebrew) One who hears. #3

Simonette (Hebrew) One who hears. #3

Simonne (Hebrew) One who hears. #8

Sinead (Hebrew) God's gracious gift. #7

Siobhan (Hebrew) Admired. #5

Sirena (Greek) A siren. #3

Sofia (Greek) Wisdom. #5

Solana (Spanish) Sunshine. #8

Solange (Latin) Alone. #1

Soledad (Latin) Alone. #6

Solenne (Spanish) Sunshine. #3

Sondra (Greek) Helper of humankind. #8

Sonia (22/4) (Greek) Wisdom. #4

Sonja (Greek) Wisdom. #5

Sonya (Greek) Wisdom. #2

Sophia (Greek) Wisdom. #5

Sophie (Greek) Wisdom. #9

Sorelle (Old French) Reddish-brown. #5

Sorrell (Old French) Reddish-brown. #9

Spring (Old English) Springtime. #2

Star (English) Star. #4

Starla (English) Star. #8

Starlene (English) Star. #4

Starlin (English) Star. #3

Starr (22/4) (English) Star. #4

Stefanie (Greek) Crowned. #7

Steffi (Greek) Crowned. #2

Stella (Latin) Star. #6

Stephanie (Greek) Crowned. #7

Stormie (English) Stormy. #9

Sue (Hebrew) Graceful lily. #9

Sugar (English) Sugar. #3

Sukey (Hebrew) Graceful lily. #9

Suki (Hebrew) Graceful lily. #6

Sula (Icelandic) Large sea bird. #8

Sumer (22/4) (Old English) Summer. #4

Sumiko (Japanese) Child of goodness. #7

Summer (English) Summer season. #8

Sunny (English) Bright, cheerful. #3

Sunshine (English) Bright, cheerful. #1

Susan (11/2) (Hebrew) Graceful lily. #2

Susanetta (Hebrew) Graceful lily. #3

Susanna (Hebrew) Graceful lily. #8

Susannah (Hebrew) Graceful lily. #7

Susanne (Hebrew) Graceful lily. #3

Susie (Hebrew) Graceful lily. #1

Suzanne (Hebrew) Graceful lily. #1

Suzie (Hebrew) Graceful lily. #8

Suzu (Japanese) Long-lived. #6

Sybil (22/4) (Greek) Prophetess. #4

Sybilla (Greek) Prophetess. #8

Sydel (Old French) From Saint Denis. #2

Sydelle (Old French) From Saint Denis. #1

Sylvia (Latin) From the forest. #7

Syndi (Greek) The moon. #8

Tabatha (Greek) Gazelle. #8

Tabby (Greek) Gazelle. #5

Tabina (Arabic) Muhammad's follower. #2

Tabitha (Greek) Gazelle. #7

Tacita (Latin) To be silent. #9

Tacy (Latin) To be silent. #4

Taffy (22/4) (Welsh) Beloved. #4

Taima (American Indian) Crash of thunder. #8

Taja (Hindi) Crown. #5

Talia (Greek) Joyful or blooming. #7

Talitha (Hebrew) Child. #8

Tallis (Old French) Woodland. #1

Tallulah (American Indian) Leaping water. #6

Tama (Japanese) Well-polished. #8

Tamara (Hebrew) Palm tree. #9

Tameko (Japanese) Child of good. #2

Tami (Japanese) People. #7

Tamiko (Japanese) People. #6

Tammy (Hebrew) Perfection. #9

Tamra (Hebrew) Palm tree. #8

Tamsin (22/4) (Hebrew) The devoted sister. #4

Tandy (Greek) Immortality. #1

Tangerine (English) Tangerine-colored. #3

Tangie (English) Tangerine-colored. #2

Tangy (22/4) (English) Tangerine-colored. #4

Tani (Greek) Giant. #8

Tania (Greek) Giant. #9

Tansey (Greek) Immortality. #3

Tansy (Greek) Immortality. #7

Tanya (Greek) Giant. #7

Tara (Gaelic) Rocky pinnacle. #4

Taree (22/4) (Japanese) Bending branch. #4

Tasha (Latin) Natal day. #4

Tatiana (Russian) Fairy queen. #3

Tatum (Old English) Cheerful. #3

Tavia (Latin) The eighth. #8

Tawny (English) Of a warm sandy color. #2

Teena (Greek) Anointed. #9

Telma (Greek) A nursling. #6

Tempest (Old French) Stormy. #8

Teodora (33/6) (Greek) Gift of God. #6

Tera (Japanese) Calm. #8

Teresa (Greek) The harvester. #5

Teresita (Greek) The harvester. #7

Tereza (Greek) The harvester. #3

Terra (Latin) The planet Earth, land. #8

Tertia (Latin) The third. #1

Tess (Greek) The harvester. #9

Tessa (Greek) The harvester. #1

Tessie (Greek) The harvester. #5

Tessy (Greek) The harvester. #7

Thalassa (Greek) From the sea. #9

Thalia (Greek) Joyful or blooming. #6

Thea (Greek) Goddess. #7

Thelma (Greek) A nursling. #5

Theodora (Greek) Gift of God. #5

Theodosia (Greek) Gift of God. #6

Theola (Greek) Gift of God. #7

Thera (Greek) Wild, untamed. #7

Theresa (Greek) The harvester. #4

Therese (Greek) The harvester. #8

Thirza (Hebrew) Pleasantness. #1

Thomasina (Hebrew) The devoted sister. #1

Thomasine (Hebrew) The devoted sister. #5

Thora (Old Norse) Thunder. #8
Thyra (Greek) Shield-bearer. #9
Tia (Spanish) Aunt. #3
Tiana (Russian) Fairy queen. #9
Tiara (22/4) (Latin) Headdress. #4
Tiena (22/4) (Greek) Anointed. #4
Tiffani (Greek) Appearance of God. #2
Tiffanie (Greek) Appearance of God. #7
Tiffany (Greek) Appearance of God. #9
Tilda (Old German) Powerful in battle. #1
Tillie (Old German) Powerful in Battle. #4
Tilly (Old German) Powerful in battle. #6
Timothea (Greek) Honoring God. #1
Tina (Greek) Anointed. #8
Tine (Greek) Anointed. #3
Tiphani (Greek) Appearance of God. #5
Tirza (Hebrew) Cypress tree. #2
Tita (Latin) Title of honor. #5
Titania (Greek) Giant. #2
Tomasine (33/6) (Hebrew) The devoted sister. #6
Tomiko (Japanese) Happiness child. #2

Toni (22/4) (Latin) Praiseworthy, without a peer. #4
Tonia (Latin) Praiseworthy, without a peer. #5
Tonya (Latin) Praiseworthy, without a peer. #3
Topaz (Latin) Topaz gem. #6
Torie (Latin) Victorious. #4
Torrie (Latin) Victorious. #4
Tory (Latin) Victorious. #6
Toshi (Japanese) Mirror reflection. #8
Tova (Hebrew) God is good. #4
Trenna (Greek) Pure maiden. #9
Tricia (33/6) (Latin) Wellborn noble. #6
Trilby (Italian) To sing with trills. #5
Trina (Greek) Pure maiden. #8
Trinidad (Latin) Triad. #7
Trinity (Latin) Triad. #7
Trish (Latin) Wellborn, noble. #2
Trisha (Latin) Wellborn, noble. #3
Trista (Latin) Melancholy. #6
Trixie (Latin) Bringer of joy. #4
Trixy (33/6) (Latin) Bringer of Joy. #6
Trude (Old German) Beloved. #5
Trudey (Old German) Beloved. #3

Trudy (Old German) Beloved. #7

Tuesday (Old English) Tuesday. #5

Tullia (Gaelic) Peaceful one. #3

Twila (Old English) Woven of a double thread. #2

Twyla (Old English) Woven of a double thread. #9

Tyne (Old English) River. #1

Tyra (Scandinavian) Of Tyr, god of battle. #1

Udele (Old English) Prosperous. #2

Udelle (Old English) Prosperous. #5

Ula (Celtic) Sea jewel. #7

Ulane (Polynesian) Cheerful. #8

Ulani (Polynesian) Cheerful. #3

Ulla (Celtic) Sea jewel. #1

Ulrica (Old German) Wolf-ruler. #1

Ulrika (Old German) Wolf-ruler. #9

Umeko (Japanese) Plum blossom child. #2

Una (Latin) Unity. #9

Unique (33/6) (Latin) The only one. #6

Unity (Old English) Unity. #8

Ursala (Latin) Little bear. #9

Ursula (Latin) Little bear. #2

Ursulina (Latin) Little bear. #7

Vala (Old German) The chosen one. #9

Valarie (Latin) Strong. #5

Valda (Old German) The ruler. #4

Valencia (Latin) Strong. #4

Valentina (Latin) Strong. #8

Valera (Latin) Strong. #5

Valeria (Latin) Strong. #5

Valerie (Latin) Strong. #9

Valora (Latin) Strong, valorous. #6

Vana (11/2) (Greek) Butterflies. #2

Vanessa (Greek) Butterflies. #9

Vanna (Greek) Butterflies. #7

Veda (Sanskrit) Wise. #5

Velda (Sanskrit) Wise. #8

Velma (Old German) Determined guardian. #8

Velvet (Old English) Velvet. #5

Venetia (Latin) From Venice. #4

Venus (Latin) Venus. #9

Vera (Latin) Truth. #1

Verda (Latin) Young, fresh. #5

Verena (Old German) Defender. #2

Verita (Latin) Truth. #3

Verity (Latin) Truth. #9

Verla (22/4) (Latin) Truth. #4

Verna (Latin) Springlike, youthful. #6

Vernice (Latin) Springlike, youthful. #4

Verona (Latin) Lady of Verona. #3

Veronica (Greek) Bringer of victory. #6

Veronika (Latin) Springlike, youthful. #5

Veronique (Greek) Bringer of victory. #9

Vesta (Latin) She who dwells. #4

Vicki (Latin) Victorious. #9

Vicky (Latin) Victorious. #7

Victoria (Latin) Victorious. #7

Vida (Sanskrit) Wise. #9

Vilette (Old French) From the village. #3

Vilma (Old German) Determined guardian. #3

Viola (Latin) Violet flower. #5

Violet (Latin) Violet flower. #2

Virginia (Latin) Virginal. #8

Viridiana (Latin) Green. #6

Vita (Latin) Alive. #7

Vitoria (Latin) Victorious. #4

Vivianne (Latin) Alive. #6

Vondra (Czech) Womanly, brave. #2

Wallis (22/4) (Old English) From Wales. #4

Wanda (Old German) The wanderer. #7

Warda (Old German) Guardian. #2

Wendy (Old German) The wanderer. #8

Wendye (Old German) The wanderer. #4

Wenona (American Indian) Firstborn daughter. #9

Wilda (22/4) (Old English) Willow. #4

Wilfreda (Old German) Resolute and peaceful. #6

Wilhelmina (Old German) Determined guardian. #7

Willa (Old German) Determined guardian. #3

Willow (Old English) Willow. #4

Wilma (22/4) (Old German) Determined guardian. #4

Wilona (Old English) Desired. #2

Wilone (33/6) (Old English) Desired. #6

Winifred (Old German) Peaceful friend. #7

Winna (African) Friend. #7

Winnie (Old German) Peaceful friend. #2

Winola (Old German) Gracious friend. #2

Winona (American Indian) Firstborn daughter. #4

Winonah (American Indian) Firstborn daughter. #3

Winter (Old English) Winter. #8

Wren (Old English) Wren. #6

Wynne (Celtic) Fair white
maiden. #9

Xandra (Greek) Helper of
humankind. #8

Xantha (Greek) Golden yellow.
#5

Xanthe (Greek) Golden yellow.
#9

Xanthippe (Greek) Yellow
horse. #5

Xaviera (Arabic) Bright. #8

Xena (Greek) Hospitable. #8

Xenia (Greek) Hospitable. #8

Xiomara (Old German) Famous
in battle. #9

Xuxa (Hebrew) Graceful lily.
#7

Xylia (Greek) Of the wood. #8

Yasmeen (Arabic) Jasmine
flower. #1

Yasmin (Arabic) Jasmine flower.
#9

Yasmine (Arabic) Jasmine
flower. #5

Yasu (Japanese) The tranquil.
#3

Yedda (Old English) The singer.
#3

Yesenia (33/6) (Arabic) Flower.
#6

Yetta (Old English) Giver. #8

Ynes (Greek) Pure, chaste. #9

Ynez (Greek) Pure, chaste. #7

Yoko (Japanese) The positive
female. #3

Yolanda (Greek) Violet flower.
#9

Yonina (33/6) (Hebrew) Dove.
#6

Yoninah (Hebrew) Dove. #5

Yoshi (Japanese) Good. #4

Yoshiko (Japanese) Good. #3

Yumiko (Japanese) Lily child.
#4

Yvette (Old French) Archer with
a yew bow. #7

Yvonne (Old French) Archer
with a yew bow. #5

Zabrina (Latin) A princess. #8

Zahara (African) Flower. #1

Zaida (Arabic) Good fortune.
#5

Zaira (Arabic) Dawning. #1

Zalika (Swahili) Wellborn. #6

Zandra (Greek) Helper of
humankind. #1

Zara (Hebrew) Of royal status.
#1

Zelda (German) Grey battle
maiden. #3

Zelia (Greek) Zealous. #8

Zena (Greek) Hospitable. #1

Zenia (Greek) Hospitable. #1

Zenobia (Arabic) Her father's
ornament. #9

Zerlina (Old German) Serene
beauty. #4

Zerlinda (Hebrew) Beautiful as
the dawn. #8

Zetta (Old English) Sixth-born. #9

Zeva (Greek) Sword. #9

Zia (Latin) kind of grain. #9

Zilla (Hebrew) Shadow. #6

Zillah (Hebrew) Shadow. #5

Zita (Latin) Little rose. #2

Zitella (Latin) A little rose. #4

Zivah (Israeli) Radiant. #3

Zoa (Greek) Life. #6

Zoe (Greek) Life. #1

Zohra (Arabic) Blossom. #5

Zola (Italian) Ball of earth. #9

Zona (Latin) A girdle or belt. #2

Zora (Greek) Dawn. #6

Zulema (Hebrew) Peace. #6

Zulima (Hebrew) Peace. #1

MASTER LIST—MALE NAMES

Aaron (22/4) (Hebrew) Lofty or exalted. #4

Abbot (Hebrew) My father is joy. #4

Abbott (Hebrew) My father is joy. #6

Abdul (Arabic) Servant of God. #4

Abdulla (Arabic) Servant of God. #8

Abdullah (Arabic) Servant of God. #7

Abe (Hebrew) My father is joy. #8

Abel (11/2) (Hebrew) Breath. #2

Abelard (Old German) Noble and resolute. #7

Abner (22/4) (Hebrew) Father of light. #4

Abraham (Hebrew) Father of the multitude. #8

Abram (Hebrew) The lofty one is father. #8

Absalom (Hebrew) Father of peace. #9

Ace (Latin) Unity. #9

Achilles (33/6) (Greek) From the river Achillios. #6

Adair (Scottish Gaelic) From the oak tree ford. #6

Adam (Hebrew) Of the red earth. #1

Adan (11/2) (Hebrew) Of the red earth. #2

Addison (Old English) Son of Adam. #3

Adiel (22/4) (Hebrew) God is an ornament. #4

Adin (Hebrew) Sensual. #1

Adlai (Hebrew) My witness. #9

Adler (22/4) (Old German) Eagle. #4

Adney (22/4) (Old English) Dweller on the island. #4

Adolf (Old German) Noble wolf. #2

Adolfo (Old German) Noble wolf. #8

Adolph (Old German) Noble wolf. #2

Adolpho (Old German) Noble wolf. #8

Adon (Hebrew) Lord God. #7

Adonis (Greek) Handsome. #8

Adriel (Hebrew) Of God's flock. #4

Adwin (Ghana) Creative. #6

Agustin (Latin) Majestic dignity. #1

Ahern (Gaelic) Horse owner. #1

Ahmed (22/4) (Arabic) Praiseworthy. #4

Ahren (Old German) The eagle. #1

Aidan (Gaelic) Warmth of the home. #2

Aiden (Gaelic) Warmth of the home. #6

Aiken (22/4) (Old English) Little Adam. #4

Akeno (Japanese) In the morning. #1

Alain (Gaelic) Handsome, cheerful. #1

Alan (Gaelic) Handsome, cheerful. #1

Alaric (Old German) Rules all. #8

Alasdair (Greek) Helper of humankind. #2

Alastair (Greek) Helper of humankind. #9

Alban (Latin) White. #3

Alben (Latin) White. #7

Alberic (Old German) Elf ruler. #5

Albern (Old English) Noble warrior. #7

Albert (22/4) (Old English) Noble and brilliant. #4

Alberto (Old English) Noble and brilliant. #1

Albie (Old English) Noble and brilliant. #2

Albin (Latin) White. #2

Albion (Celtic) White cliffs. #8

Alby (Latin) White. #4

Alcott (Old English) From the cottage. #8

Alden (Old English) Old wise protector. #9

Aldis (Old German) Old and wise. #9

Aldo (Latin) Elder. #5

Aldous (Old German) Old and wise. #9

Aldrich (Old English) Old wise ruler. #1

Alec (Greek) Helper of humankind. #3

Alejandro (Greek) Helper of humankind. #8

Alessandro (Greek) Helper of humankind. #9

Alexander (Greek) Helper of humankind. #3

Alexandre (Greek) Helper of humankind. #3

Alexandro (Greek) Helper of humankind. #4

Alexio (Greek) Helper of humankind. #3

Alfonso (Old German) Noble and eager. #1

Alford (Old English) The old ford. #2

Alfred (Old English) Good counselor. #1

Alfredo (Old English) Good counselor. #7

Alger (Old German) Noble spearman. #7

Algernon (Old French) Mustached or bearded. #5

Alick (Greek) Helper of humankind. #9

Alistair (Greek) Helper of humankind. #8

Alister (Greek) Helper of humankind. #3

Allain (22/4) (Gaelic) Handsome, cheerful. #4

Allan (Gaelic) Handsome, cheerful. #4

Allard (Old English) Noble and brave. #3

Allen (Gaelic) Handsome, cheerful. #8

Allister (33/6) (Greek) Helper of humankind. #6

Allon (Gaelic) Handsome and cheerful. #9

Allun (Gaelic) Handsome, cheerful. #6

Allyn (Gaelic) Handsome, cheerful. #1

Alonso (22/4) (Old German) Noble and eager. #4

Alonzo (Old German) Noble and eager. #2

Aloysius (Old German) Famous in war. #4

Alphonse (Old German) Noble and eager. #9

Altman (Old German) Wise old man. #7

Alton (English) From the old town. #8

Alvah (Latin) White, fair. #8

Alvan (Old German) Beloved by all. #5

Alvaro (Old German) All wise, prudent. #6

Alvin (22/4) (Old German) Beloved by all. #4

Alvino (Old German) Beloved by all. #1

Alvis (Old English) All-knowing. #9

Alvyn (Old German) Beloved by all. #2

Alwin (Old German) Beloved by all. #5

Alyn (Gaelic) Handsome, cheerful. #7

Amadeo (Latin) Love of God. #3

Amadeus (Latin) Love of God. #1

Ambrose (Greek) Immortal. #1

Ambrosius (Greek) Immortal. #9

Amery (Old German) Industrious ruler. #8

Ames (11/2) (Old French) Friend. #2

Amiel (22/4) (Hebrew) Lord of my people. #4

Amir (Arabic) Prince. #5

Amon (Hebrew) Trustworthy, faithful. #7

Amory (Old German) Industrious ruler. #9

Amos (Hebrew) Burden. #3

Anatole (Greek) From the east. #5

Anders (Greek) Strong and manly. #7

Anderson (Greek) Strong and manly. #9

Andre (Greek) Strong and manly. #6

Andres (Greek) Strong and manly. #7

Andrew (Greek) Strong and manly. #2

Angelo (Greek) Heavenly messenger. #9

Angus (Celtic) Unique strength. #8

Annell (22/4) (Celtic) Beloved. #4

Ansel (Old French) Adherent of a nobleman. #6

Ansell (Old French) Adherent of a nobleman. #9

Anselm (Old German) Divine warrior. #1

Anselmo (Old German) Divine warrior. #7

Ansley (22/4) (Old English) From Ann's meadow. #4

Anson (Old German) Of divine origin. #9

Anthony (Latin) Praiseworthy, without a peer. #7

Antoine (33/6) (Latin) Praiseworthy, without a peer. #6

Anton (Latin) Praiseworthy, without a peer. #1

Antonio (Latin) Praiseworthy, without a peer. #7

Antony (Latin) Praiseworthy, without a peer. #8

Apollo (Greek) Manly. #8

Ara (11/2) (Arabic) Rainmaker. #2

Aram (Assyrian) High, exalted. #6

Archer (Old English) Bowman. #8

Archibald (Old German) Genuinely bold. #4

Archie (Old German) Genuinely bold. #8

Ardley (Old English) From the domestic meadow. #2

Argus (Greek) Watchful. #3

Aric (22/4) (Old English) Sacred ruler. #4

Arick (Old English) Sacred ruler. #6

Aristotle (Greek) The best. #2

Arlen (Gaelic) A pledge. #5

Arlo (Spanish) The barberry. #1

Armand (Old German) Army man. #6

Armando (Old German) Army man. #3

Armon (Hebrew) High place. #7

Armstrong (Old English) Strong arm. #8

Army (Old German) Army man. #3

Arnall (22/4) (Old German) Strong as an eagle. #4

Arnan (Hebrew) Quick, joyful. #3

Arnaud (Old German) Strong as an eagle. #5

Arne (Old German) Strong as an eagle. #2

Arney (Old German) Strong as an eagle. #9

Arnie (Old German) Strong as an eagle. #2

Arno (Old German) Strong as an eagle. #3

Arnold (Old German) Strong as an eagle. #1

Arnoldo (Old German) Strong as an eagle. #7

Arsenio (Greek) Potent. #9

Arsenius (Greek) Potent. #7

Art (Celtic-Welsh) Noble, bear-hero. #3

Artemus (Greek) Gift of Artemis. #7

Arthur (Celtic-Welsh) Noble, bear-hero. #5

Artie (Celtic-Welsh) Noble, bear-hero. #8

Arvin (Old German) People's friend. #1

Asa (Hebrew) Physician. #3

Ase (Hebrew) Physician. #7

Ashby (Old Norse) From the ash tree farm. #1

Asher (Hebrew) Divinely gifted. #6

Ashford (Old English) From the ash tree ford. #8

Atherton (Old English) Dweller at the spring farm. #2

Atwell (Old English) From the spring. #1

Auberon (Old German) Elf ruler. #4

Aubert (22/4) (Old English) Noble and brilliant. #4

Aubin (Old French) The blond one. #2

Audel (Old English) Old friend. #7

Auden (Old English) Old friend. #9

Audie (22/4) (Old English) Old friend. #4

Audric (Old English) Wise old ruler. #2

Audwin (Old German) Noble friend. #9

August (Latin) Majestic dignity. #8

Augustine (Latin) Majestic dignity. #9

Augustus (Latin) Majestic dignity. #3

Aurelio (Latin) Golden. #9

Aurelius (Latin) Golden. #7

Avenall (22/4) (Old French) Dweller in the oat field. #4

Averill (Latin) Forthcoming. #7

Avram (Hebrew) The lofty one is father. #1

Axel (Hebrew) Father of peace. #6

Axton (Old English) From the peaceful town. #2

Ayers (Old English) Heir to a fortune. #5

Bainbridge (Old English) Bridge over white water. #8

Baird (Gaelic) Ballad singer. #7

Baldemar (Old German) Bold and famous prince. #2

Baldric (Old German) Princely ruler. #4

Baldwin (Old German) Bold friend. #2

Balfour (Scottish Gaelic) Pasture land. #3

Bancroft (Old English) From the bean field. #7

Barak (Hebrew) Flash of lightning. #6

Baran (Russian) Forceful, virile. #9

Barclay (Old English) From the birch tree meadow. #8

Bard (Gaelic) Poet and singer. #7

Barlow (Old English) From the bare hill. #8

Barnabas (22/4) (Greek) Son of prophecy. #4

Barnaby (Greek) Son of prophecy. #9

Barnett (Old English) Nobleman. #8

Barney (Greek) Son of prophecy. #2

Baron (Old English) Nobleman, baron. #5

Barret (Old English) Glorious raven. #1

Barrett (Old German) Mighty as a bear. #3

Barric (33/6) (Hebrew) Flash of lightning. #6

Barry (Celtic) One whose intellect is sharp. #1

Bart (Hebrew) Son of a farmer. #5

Bartholomew (Hebrew) Son of a farmer. #6

Bartley (Old English) Bartholomew's meadow. #2

Barton (Old English) From the barley farm. #7

Bartram (Old English) Brilliant raven. #1

Baruch (Hebrew) Blessed. #8

Basil (Latin) Kingly, magnificent. #7

Basir (22/4) (Turkish) Intelligent and discerning. #4

Baxter (Old English) Baker. #7

Bayard (Old English) Having reddish-brown hair. #6

Beacher (33/6) (Old English) One who lives by oak trees. #6

Beal (11/2) (Old French) Handsome. #2

Beale (Old French) Handsome. #7

Beau (11/2) (Old French) Handsome. #2

Beaumont (Old French) Beautiful mountain. #1

Beauregard (Old French) Beautiful expression. #1

Beck (Old English) A brook. #3

Beecher (Old English) One who lives by oak trees. #1

Belden (Old English) Dweller in the beautiful glen. #6

Bellamy (Old French) Handsome friend. #7

Ben (Hebrew) Son of my right hand. #3

Benedict (Latin) Blessed. #8

Benito (Hebrew) Son of my right hand. #2

Benjamin (Hebrew) Son of my right hand. #5

Benjiro (Japanese) Enjoy peace. #1

Benjy (Hebrew) Son of my right hand. #2

Bennett (Latin) Blessed. #8

Benny (Hebrew) Son of my right hand. #6

Benson (Old English) Son of Benjamin. #6

Bentley (Old English) From the bent grass meadow. #2

Benton (Old English) From the bent grass town. #7

Beresford (Old English) From the barley wood. #2

Berkley (33/6) (Old English) From the birch meadow. #6

Berl (Old English) Cupbearer. #1

Berle (Old English) Cupbearer. #6

Berlyn (Old English) Son of Burl. #4

Bern (Old German) Brave as a bear. #3

Bernard (Old German) Brave as a bear. #8

Bernardo (Old German) Brave as a bear. #5

Berne (Old German) Brave as a bear. #8

Bernie (Latin) Brave as a bear. #8

Bert (Old English) Bright. #9

Berthold (Old German) Brilliant ruler. #3

Bertie (Old English) Bright. #5

Bertram (Old English) Brilliant raven. #5

Bertrand (Old English) Brilliant raven. #1

Beto (Old English) Noble and brilliant. #6

Bevan (Celtic) Youthful warrior. #8

Bill (Old German) Determined guardian. #8

Bing (Old German) Kettle-shaped hollow. #5

Birch (Old English) Birch tree. #4

Bishop (33/6) (Old English) Bishop. #6

Bjorn (Old German) Bear. #5

Blade (Old English) Knife, sword. #6

Blakeley (Old English) From the black meadow. #1

Blanco (Spanish) White, fair. #2

Blase (Latin) Stammerer. #3

Boaz (Hebrew) In the Lord is strength. #8

Bob (Old English) Bright fame. #1

Boden (22/4) (Old English) Herald, messenger. #4

Bogart (Old German) Strong as a bow. #9

Bond (Old English) Tiller of the soil. #8

Boniface (Latin) One who does good. #1

Boone (Old French) Good. #6

Booth (Old English) From the hut. #6

Borak (Arabic) The lightning. #2

Borden (Old English) From the valley of the boar. #4

Borg (Old Norse) Castle dweller. #6

Boris (Slavic) Battler, warrior. #9

Bowen (Welsh) Son of Owen. #5

Bowie (Gaelic) Yellow-haired. #9

Boyce (Old French) From the woodland. #5

Boyd (Gaelic) Blond. #1

Boyne (Gaelic) A rare person. #7

Brad (Old English) From the broad meadow. #7

Braddock (Old English) Broad-spreading oak. #4

Braden (Old English) From the broad valley. #8

Bradford (Old English) From the broad river crossing. #5

Bradlee (Old English) From the broad meadow. #2

Bradleigh (Old English) From the broad meadow. #3

Bradley (Old English) From the broad meadow. #4

Bradshaw (Old English) From the large virginal forest. #4

Brady (Old English) From the broad island. #5

Bram (Gaelic) Raven. #7

Bramwell (Old English) At Bram's well. #5

Brand (Old English) Firebrand. #3

Brando (Old English) Firebrand. #9

Brandon (Old English) From the beacon hill. #5

Brandt (Old English) Firebrand. #5

Brant ((Old English) Firebrand. #1

Braun (Old German) Brown-haired. #2

Braxton (Old English) From Braxton. #4

Breck (Gaelic) Freckled. #3

Brendan (Gaelic) Little raven. #4

Brennan (Gaelic) Little raven. #5

Brent (Old English) Steep hill. #5

Brenton (Old English) From the town on the steep hill. #7

Bret (Celtic) Native of Brittany. #9

Brewster (Old English) Brewer. #2

Brexton (Gaelic) Freckled. #8

Brian (Celtic) Strength with virtue and honor. #8

Briant (Celtic) Strength with virtue and honor. #1

Brice (Celtic) Quick. #1

Brick (Old English) Bridge. #7

Brickman (Old English) Bridge man. #8

Brien (Celtic) Strength with virtue and honor. #3

Brigham (Old English) From the enclosed bridge. #4

Brock (22/4) (Old English) Badger. #4

Broderick (Old English) From the broad ridge. #4

Brodie (Gaelic) A ditch. #8

Brodrick (Old English) From the bridge road. #8

Brody (Gaelic) A ditch. #1

Bronson (Old English) Son of the dark-skinned one. #7

Bruce (22/4) (French) From the thicket. #4

Bruno (Old German) Brown-haired. #7

Bryan (Celtic) Strength with virtue and honor. #6

Bryant (Celtic) Strength with virtue and honor. #8

Bryce (Celtic) Quick. #8

Bryon (Celtic) Strength with virtue and honor. #2

Buchanan (Scottish Gaelic) House of the clergy. #1

Buck (Old English) Buck deer. #1

Bud (Old English) Herald, messenger. #9

Budd (Old English) Herald, messenger. #4

Buddie (Old English) Herald, messenger. #9

Buddy (Old English) Herald, messenger. #2

Buell (Old German) Hill dweller. #7

Burdett (Old French) Little shield. #9

Burdette (Old French) Little shield. #5

Burdon (Old English) From the castle on the hill. #2

Burgess (Old English) Citizen of the fortified town. #1

Burke (Old French) Dweller at the fortress. #3

Burl (Old English) Cupbearer. #8

Burley (Old English) From the castle by the meadow. #2

Burnard (33/6) (Old German) Brave as a bear. #6

Burne (Old English) Brook. #6

Burnett (Old English) Brown complexion. #1

Burney (Old English) Brook. #4

Burt (Old English) From the fortress. #7

Burton (Old English) From the fortress. #9

Butch (Old English) Bright. #9

Byrne (Old English) From the brook. #1

Byron (Old French) From the cottage. #2

Cabe (11/2) (Old English) Ropemaker. #2

Cable (Old English) Ropemaker. #5

Cadell (Celtic) Battle spirit. #1

Caden (Celtic) Battle spirit. #9

Caelan (Gaelic) Powerful in battle. #9

Caesar (Latin) Long-haired, emperor. #2

Cain (Hebrew) Possession or possessed. #9

Caine (Hebrew) Possession or possessed. #5

Cal (Latin) Bald. #7

Calder (Old English) The stream. #7

Caldwell (Old English) Cold spring. #9

Cale (Hebrew) Spontaneous. #3

Caleb (Hebrew) Spontaneous. #5

Calhoun (Celtic) From the narrow forest. #2

Callum (Gaelic) Mild, gentle. #8

Calvert (Old English) Herdsman. #9

Calvin (Latin) Bald. #7

Camden (22/4) (Scottish Gaelic) From the winding valley. #4

Camillo (Latin) Young ceremonial attendant. #2

Camilo (Latin) Young ceremonial attendant. #8

Campbell (Scottish Gaelic) Crooked mouth. #1

Cantrell (Latin) Singer. #4

Canute (Old Norse) Knot. #1

Caradoc (Celtic) Beloved. #9

Caradock (Celtic) Beloved. #2

Carl (Old German) Strong and masculine. #7

Carleton (Old English) Farmer's town. #7

Carlin (Gaelic) Little champion. #3

Carlisle (Old English) From the fortified town. #7

Carlo (22/4) (Old German) Strong and masculine. #4

Carlos (Old German) Strong and masculine. #5

Carlton (Old English) Farmer's town. #2

Carlyle (Old English) From the fortified town. #4

Carmelo (Latin) Song. #4

Carmine (Latin) Song. #9

Carne (Gaelic) Victorious. #5

Carnell (Old English) Defender of the castle. #2

Carney (Gaelic) Victorious. #3

Carr (22/4) (Old Norse) He who dwells by the marsh. #4

Carrick (Scottish) From the rugged hills. #9

Carson (Old English) Son of the marsh dwellers. #7

Carsten (Greek) Anointed. #8

Carter (Old English) Cart driver. #2

Carver (Old English) Woodcarver. #4

Cash (Latin) Vain. #4

Cashman (Latin) Vain. #5

Casimir (Slavic) Proclamation of peace. #9

Caspar (22/4) (Persian) Treasurer. #4

Casper (Persian) Treasurer. #8

Cassius (Latin) Vain. #1

Castor (22/4) (Greek) Industrious. #4

Cathmor (33/6) (Gaelic) Great warrior. #6

Cecil (Latin) Dim-sighted or blind. #5

Cedric (33/6) (Old English) Battle chieftain. #6

Cesar (Latin) Long-haired, emperor. #1

Cesare (Latin) Long-haired, emperor. #6

Chace (Old French) Hunter. #2

Chad (Old English) Warlike. #7

Chadwick (Old English) From the warrior's town. #8

Chaim (Hebrew) Life. #7

Chalmers (Scottish Gaelic) Son of the overseer. #7

Chance (English) Fortune, a gamble. #7

Chancellor (Old English) King's counselor. #1

Chaney (French) Oak wood. #2

Chapman (Old English) Merchant. #2

Charles (Old German) Strong and masculine. #3

Charley (Old German) Strong and masculine. #9

Charlton (Old English) Farmer's town. #1

Chase (Old French) Hunter. #9

Chasen (Old French) Hunter. #5

Chauncey (Old English) Chancellor. #8

Che (Hebrew) He shall add. #7

Chen (Chinese) Great. #3

Cheney (33/6) (Old French) Dweller in the oak forest. #6

Cherokee (American Indian) People of a different speech. #7

Chesney (Old French) Dweller in the oak forest. #7

Chester (33/6) (Old English) From the fortified camp. #6

Chet (Old English) From the fortified camp. #9

Cheval (Old French) Horseman, knight. #6

Chevy (Old French) Horseman, knight. #9

Chic (Old German) Strong and masculine. #5

Chick (Old German) Strong and masculine. #7

Chico (Spanish) Small boy. #2

Chilton (Old English) From the farm by the spring. #9

Christian (Greek) Anointed. #2

Christopher (Greek) Bearer of Christ. #4

Chuck (Old German) Strong and masculine. #1

Churchill (Old English) By the church on the hill. #4

Cicero (Latin) Chickpea. #8

Claiborn (Old English) Born of clay. #2

Clancy (22/4) (Celtic) Redheaded. #4

Clarence (Latin) Bright, famous. #7

Clark (Old French) Scholar. #9

Clarke (Old French) Scholar. #5

Claud (Latin) Lame. #5

Claude (Latin) Lame. #1

Claudio (Latin) Lame. #2

Claus (11/2) (Latin) Lame. #2

Clay (Old English) From the place of clay. #5

Clayborne (Old English) Born of clay. #5

Clayton (Old English) From the town built on clay. #9

Cleary (Gaelic) The scholar. #1

Cleavon (Old English) From the cliff. #9

Clemens (Latin) Merciful. #8

Clement (Latin) Merciful. #9

Cleon (22/4) (Greek) Famous. #4

Cletis (Greek) Famous. #5

Cletus (Greek) Famous. #8

Cleve (Old English) From the cliff-land. #2

Cleveland (33/6) (Old English) From the cliff-land. #6

Cliff (Old English) From the cliff. #9

Clifford (Old English) From the cliff-ford. #1

Cliffton (Old English) From the cliff-town. #4

Clifton (Old English) From the cliff-town. #7

Clint (22/4) (Old English) From the hillside. #4

Clinton (33/6) (Old English) From the hillside town. #6

Clive (Old English) From the cliff. #6

Clovis (Latin) Renowned fighter. #8

Clyde (22/4) (Celtic) Heard from afar. #4

Colbert (Old English) Brilliant seafarer. #3

Colby (Old English) From the dark farm. #3

Cole (Greek) The people's victory. #8

Coleman (Old English) Nicholas's man. #9

Colin (Gaelic) Strong and virile. #8

Collier (Old English) Charcoal merchant, miner. #2

Collin (Gaelic) Strong and virile. #2

Colm (Latin) Dove. #7

Colt (English) Young horse. #5

Colter (Old English) The colt herder. #1

Colton (Old English) From the coal town. #7

Colum (Latin) Dove. #1

Conan (Celtic) High and mighty. #2

Conlan (Gaelic) Hero. #5

Connal (Celtic) Mighty, valorous. #5

Connell (Celtic) Mighty, valorous. #3

Connor (Gaelic) Wise man. #7

Conrad (Old German) Bold counselor. #1

Conroy (Gaelic) Wise one. #9

Constant (Latin) Firmness, constancy. #7

Constantine (Latin) Firmness, constancy. #8

Conway (Gaelic) Hound of the plain. #9

Cooper (Old English) Barrelmaker. #9

Corbett (Latin) Raven. #2

Corbin (Latin) Raven. #7

Corcoran (Gaelic) Reddish complexion. #6

Cordell (33/6) (Old French) Ropemaker. #6

Cormick (Gaelic) Charioteer. #9

Cornel (Latin) Yellow, horn-colored. #4

Cornelius (Latin) Yellow, horn-colored. #8

Cornell (Latin) Yellow, horn-colored. #7

Corrigan (Celtic) Spearman. #4

Cort (Old German) Bold. #2

Cortez (33/6) (Spanish) Courteous. #6

Corwin (Old French) Friend of the heart. #1

Cosmo (Greek) The universe, well-ordered. #2

Courtland (Old English) From the court land. #9

Cowan (Gaelic) Hillside hollow. #2

Coyle (Celtic) Leader in battle. #6

Craig (Scottish Gaelic) From near the crag. #2

Crandall (Old English) From the crane valley. #2

Cranley (33/6) (Old English) From the crane meadow. #6

Crawford (Old English) From the ford of the crow. #7

Creed (Latin) Belief, guiding principle. #8

Creighton (Old English) From the town near the creek. #9

Crispian (Latin) Curly-haired. #8

Crispin (Latin) Curly-haired. #7

Cromwell (Old English) Dweller by the winding brook. #2

Crosby (Scandinavian) From the shrine of the cross. #1

Cross (Scandinavian) From the shrine of the cross. #2

Culbert (Old English) Brilliant seafarer. #9

Cullen (22/4) (Gaelic) Handsome. #4

Culver (Old English) Dove. #9

Curran (Gaelic) Champion or hero. #3

Curt (Old French) Courteous. #8

Curtis (Old French) Courteous. #9

Curtiss (Old French) Courteous. #1

Cutler (Old English) Knifemaker. #7

Cynric (Old English) From the royal line of kings. #9

Cyril (Greek) Master, ruler. #4

Cyrus (Old Persian) The sun. #5

Dafydd (Hebrew) Beloved. #8

Dag (Scandinavian) Day, brightness. #3

Dagwood (33/6) (Old English) Forest of the shining one. #6

Daley (Gaelic) Assembly, gathering. #2

Dalston (22/4) (Old English) From Daegal's town. #4

Dalton (Old English) From the estate in the valley. #3

Daly (Gaelic) Assembly, gathering. #6

Damario (Greek) Gentle. #7

Damian (Greek) Constant. #6

Damien (Greek) Constant. #1

Damon (Greek) Constant. #2

Dan (Hebrew) God is my judge. #1

Danby (Old Norse) From the Danish settlement. #1

Dane (Old English) From Denmark. #6

Dante (Latin) Lasting. #8

Dare (Gaelic) Great. #1

Dario (Greek) Wealthy. #2

Darius (Greek) Wealthy. #9

Darnell (Old English) From the hidden place. #3

Darrel (Old French) Beloved. #4

Darrell (Old French) Beloved. #7

Darrick (Scottish Gaelic) Strong, oak-hearted. #1

Darrin (Greek) Gift. #1

Darvell (Old English) Town of eagles. #2

Darwin (33/6) (Old English) Beloved friend. #6

Dave (Hebrew) Beloved. #5

Davey (Hebrew) Beloved. #3

David (22/4) (Hebrew) Beloved. #4

Davie (Hebrew) Beloved. #5

Davis (Old English) Son of David. #1

Davy (Hebrew) Beloved. #7

Dayan (Hebrew) Judge. #9

Deacon (Greek) Servant. #6

Dean (Old English) From the valley. #6

Delaney (Gaelic) Descendant of the challenger. #3

Delano (Old French) Of the night. #6

Delbert (Old English) Bright as day. #3

Delfino (Greek) Calmness. #2

Delmar (Old French) From the sea. #8

Delmer (Old French) From the sea. #3

Delmon (Old English) Of the mountain. #9

Delmont (Old English) Of the mountain. #2

Delmore (Old French) From the sea. #9

Delroy (Old French) Of the king. #7

Delton (Old English) From the town in the valley. #7

Delvin (Old English) Godly friend. #3

Delwin (Old English) Proud friend. #4

Delwyn (Old English) Proud friend. #2

Demetrius (Greek) From the fertile land. #6

Dempsey (33/6) (Gaelic) Proud. #6

Denis (Greek) God of wine. #6

Denman (Old English) Resident in the valley. #6

Dennis (Greek) God of wine. #2

Denton (Old English) Valley town. #9

Denver (Old English) Green valley. #5

Denzel (English) From Denzel in Cornwall. #3

Denzil (Greek) God of wine. #7

Denzo (Japanese) Discreet. #1

Derek (Old German) Leader of the people. #7

Derick (Old German) Leader of the people. #5

Dermot (Gaelic) Free from
envy. #3

Derrek (Old German) Leader of
the people. #7

Derrick (Old German) Leader of
the people. #5

Derry (Gaelic) Great. #7

Derward (Old English)
Gatekeeper. #1

Deryck (Old German) Leader of
the people. #3

Desi (Latin) Yearning, sorrow.
#1

Desiderus (Latin) Yearning,
sorrow. #5

Desmond (Gaelic) Man from
the south. #2

Destin (Old French) Destiny.
#8

Devlin (Gaelic) Fierce valor. #3

Dewey (Hebrew) Beloved. #8

Dewitt (Flemish) White or
blond. #9

Dexter (Latin) Right-handed.
#4

Diccon (Old German) Wealthy,
powerful. #3

Dick (Old German) Wealthy,
powerful. #9

Diego (Hebrew) The supplanter.
#4

Dietrich (Old German) Ruler of
the people. #4

Dillon (Gaelic) Faithful. #3

Dimitri (Greek) From the fertile
land. #1

Dinsmore (Gaelic) From the
fortified hill. #7

Dion (Greek) God of wine. #6

Dionisio (Greek) God of wine.
#4

Dirk (Old German) Leader of
the people. #6

Dolan (Celtic) Dark, bold. #1

Dolf (Old German) Noble Wolf.
#1

Dolph (Old German) Noble
Wolf. #1

Domingo (Latin) The Lord's
day. #5

Dominic (Latin) Belonging to
the Lord. #4

Dominick (Latin) Belonging to
the Lord. #6

Don (Gaelic) World ruler. #6

Donahue (Gaelic) Brown-haired
fighter. #5

Donald (Gaelic) World ruler.
#5

Donnell (Gaelic) World ruler.
#4

Donnelly (Gaelic) Brave dark
man. #2

Donovan (Gaelic) Dark warrior.
#4

Doug (Scottish Gaelic) From the
dark water. #2

Dougal (Celtic) From the dark
stream. #6

Douglas (Scottish Gaelic) From the dark water. #7

Dov (Hebrew) Beloved. #5

Dow (Gaelic) Black-haired. #6

Doyle (Gaelic) Dark stranger. #7

Drake (Old English) Dragon. #3

Druce (Celtic) Son of Drew. #6

Drury (Old French) Beloved friend. #5

Dryden (Old English) From the dry valley. #7

Duane (Gaelic) Dark. #9

Duarte (Old English) Prosperous guardian. #6

Dudley (Old English) From the people's meadow. #8

Duff (Celtic) Dark. #1

Duffy (Celtic) Dark. #8

Dugal (Celtic) From the dark stream. #9

Dugald (22/4) (Celtic) From the dark stream. #4

Dugan (Gaelic) Dark. #2

Duke (Old French) Leader, duke. #5

Dumont (Old French) From the mountain. #6

Duncan (Scottish Gaelic) Dark-skinned warrior. #3

Dunham (Celtic) Dark man. #7

Dunstan (Old English) From the brown stone hill. #3

Durand (Latin) Enduring. #8

Durant (Latin) Enduring. #6

Durward (Old English) Gatekeeper, door warden. #8

Dustin (Old German) Valiant fighter. #6

Duval (Old French) From the valley. #6

Dwayne (Gaelic) Dark. #9

Dwight (Old English) A cutting or clearing. #8

Dylan (Welsh) Man from the sea. #2

Eagan (Celtic) Fiery, forceful. #1

Eamon (Old English) Prosperous guardian. #3

Eamonn (Old English) Prosperous guardian. #8

Earl (Old English) Nobleman. #9

Earle (Old English) Nobleman. #5

Easton (Old English) From East town. #2

Eaton (Old English) From the estate on the river. #1

Eben (Hebrew) Stone. #8

Ebenezer (Hebrew) Stone of help. #8

Edan (Celtic) Flame. #6

Eddie (Old English) Prosperous guardian. #9

Eddy (Old English) Prosperous guardian. #2

Eder (Hebrew) Flock. #5

Edgar (Old English) Prosperous
spearman. #8

Edison (Old English) Son of
Edward. #3

Edmond (Old English)
Prosperous protector. #1

Edmund (Old English)
Prosperous protector. #7

Edric (Old English) Power and
good fortune. #3

Edrick (Old English) Power and
good fortune. #5

Edsel (Old English) Rich man's
house. #9

Eduardo (Old English)
Prosperous guardian. #5

Edwald (22/4) (Old English)
Prosperous ruler. #4

Edward (Old English)
Prosperous guardian. #1

Edwin (Old English) Rich friend.
#1

Efrain (Hebrew) Very fruitful.
#8

Efrem (Hebrew) Very fruitful.
#2

Efron (Hebrew) Young stag. #4

Egan (Gaelic) Ardent, fiery. #9

Egon (Gaelic) Ardent, fiery. #5

Einar (Old Norse) Warrior chief.
#2

Ekon (Nigerian) Strong. #9

Elbert (Old German) Bright,
famous. #8

Elden (22/4) (Old English) Wise
protector. #4

Eldon (Old English) Wise
protector. #5

Eldred (Old English) Sage
counsel. #3

Eldredge (Old English) Old
wise ruler. #6

Eldridge (old English) Old wise
ruler. #1

Eleazar (Hebrew) God has
helped. #5

Elford (33/6) (Old English) From
the alder tree ford. #6

Elgar (Old German) Sword. #7

Elgin (Celtic) Noble, white. #2

Eli (Hebrew) Jehovah. #8

Elias (Hebrew) Jehovah is God.
#1

Elihu (Hebrew) My God is He.
#1

Elijah (Hebrew) Jehovah is God.
#9

Ellard (Old German) Nobly
brave. #7

Ellery (Old English) From the
elder tree island. #5

Elliot (Hebrew) Jehovah is God.
#1

Elliott (Hebrew) Jehovah is
God. #3

Ellis (Hebrew) Jehovah is God.
#3

Ellison (Old English) Son of
Ellis. #5

Ellsworth (Old English) Nobleman's estate. #6

Elmer (Old English) Noble, famous. #8

Elmo (Italian) Helmet. #9

Eloy (Latin) Chosen one. #3

Elroy (Old French) The King. #3

Elson (Old English) Son of Elias. #2

Elston (22/4) (Old English) Nobleman's town. #4

Elsworth (Old English) Nobleman's estate. #3

Elton (Old English) From the old town. #3

Elvern (Latin) Spring, greening. #4

Elvin (Old English) Elf-friend. #8

Elvis (22/4) (Norse) The prince of wisdom. #4

Elvy (Old English) Elfin warrior. #1

Elwin (Old English) Elf-friend. #9

Elwood (Old English) From the old forest. #2

Ely (Hebrew) Jehovah. #6

Emanuel (Hebrew) God is with us. #8

Emerick (Old German) Industrious ruler. #1

Emerson (Old German) Son of the industrious ruler. #8

Emery (Old German) Industrious ruler. #3

Emerys (Greek) Immortal. #4

Emil (Latin) Flattering. #3

Emilio (Latin) Flattering. #9

Emlyn (Latin) Flattering. #6

Emmanuel (Hebrew) God is with us. #3

Emmet (Old German) Industrious ruler. #2

Emmett (22/4) (Old German) Industrious ruler. #4

Emory (Old German) Industrious ruler. #4

Emrys (Greek) Immortal. #8

Engelbert (Old German) Bright as an angel. #7

Ennis (Gaelic) The only choice. #7

Enoch (Hebrew) Experienced, dedicated. #9

Enos (Hebrew) Man. #8

Enrico (Old German) Lord of the manor. #1

Enrique (Old German) Lord of the manor. #8

Enzo (Latin) Crowned with laurels. #6

Ephraim (Hebrew) Very fruitful. #7

Erasmus (Greek) Lovable. #6

Erastus (22/4) (Greek) Beloved. #4

Erhard (Old German) Strong honor. #9

Eric (Old Norse) Ever-powerful, ever-ruler. #8

Erick (Old Norse) Ever-powerful, ever-ruler. #1

Erik (Old Norse) Ever-powerful, ever-ruler. #7

Erle (22/4) (Old English) Nobleman. #4

Erling (Old English) Son of a nobleman. #2

Ernest (Old English) Earnest. #9

Ernesto (33/6) (Old English) Earnest. #6

Errol (Old English) Nobleman. #5

Erroll (Old English) Nobleman. #8

Erskine (Scottish Gaelic) From the height of the cliff. #9

Ervin (Old English) Sea friend. #5

Erwin (33/6) (Old English) Sea friend. #6

Esau (Hebrew) Hairy. #1

Esmond (Old English) Gracious protector. #7

Esteban (Greek) Crowned. #3

Estefan (Greek) Crowned. #7

Ethan (Hebrew) Firm or strong. #3

Etienne (Greek) Crowned. #9

Eugene (Greek) Wellborn noble. #3

Eustace (Greek) Steadfast. #2

Evan (Gaelic) Wellborn young warrior. #6

Even (Gaelic) Wellborn young warrior. #1

Everard (Old English) Hardy, brave. #1

Everett (Old English) Strong as a boar. #5

Everley (Old English) Field of the wild boar. #2

Evian (Hebrew) God's gracious gift. #6

Evin (Gaelic) Wellborn young warrior. #5

Ewald (Old English) Law, powerful. #9

Ewan (Gaelic) Wellborn young warrior. #7

Ewen (Gaelic) Wellborn young warrior. #2

Ewing (Old English) Friend of the law. #4

Ezekiel (Hebrew) Strength of God. #1

Ezra (Hebrew) Helper. #5

Fabian (Latin) Bean grower. #6

Fabiano (Latin) Bean grower. #3

Fabio (Latin) Bean grower. #6

Fabion (Latin) Bean grower. #2

Fabrice (Latin) Craftsman. #8

Fabricio (Latin) Craftsman. #9

Fabrizio (Latin) Craftsman. #5

Fagan (Gaelic) The fiery one. #2

Faine (Old English) Good-natured. #8

Fairfax (Old English) Fair-haired. #2

Fairleigh (Old English) From the sheep meadow. #3

Falco (Latin) Falcon. #1

Falcon (Latin) Falcon. #6

Falkner (Old English) Trainer of falcons. #4

Fane (Old English) Good-natured. #8

Faraji (Swahili) Consolation. #9

Farand (Old German) Pleasant and attractive. #8

Farid (Arabic) Exceptional, unequaled. #2

Farleigh (Old English) From the sheep meadow. #3

Farley (Old English) From the sheep meadow. #4

Farold (Old English) Mighty traveler. #2

Farquhar (Gaelic) Friendly man. #9

Farrell (Gaelic) Most valorous. #9

Farris (Old English) Iron-strong. #8

Fausto (Latin) Fortunate. #1

Favian (Latin) Bean grower. #8

Favio (Latin) Bean grower. #8

Feliciano (Latin) Joyous, fortunate. #2

Felipe (Greek) Lover of horses. #8

Felippe (Greek) Lover of horses. #6

Felix (Latin) Joyous, fortunate. #2

Felton (Old English) From the estate on the meadow. #9

Fenton (Old English) From the marshland farm. #2

Ferdinand (Old German) World-daring. #3

Fergie (Gaelic) The best choice. #5

Fergus (Gaelic) The best choice. #4

Ferguson (Gaelic) The best choice. #6

Fermin (Latin) Energetic. #2

Fernald (33/6) (Old English) From the fern slope. #6

Fernando (Old German) World-daring. #5

Ferris (Old English) Iron-strong. #3

Ferron (Old English) Ironworker. #4

Fidel (Latin) Faithful. #9

Fielding (Old English) From the field. #3

Filbert (Old English) Very brilliant. #9

Filip (Greek) Lover of horses. #7

Filmer (Old English) Very famous. #9

Filmore (Old English) Very famous. #6

Findlay (Gaelic) Little fair-haired soldier. #8

Findley (Gaelic) Little fair-haired soldier. #3

Finlay (Gaelic) Little fair-haired soldier. #4

Finley (Gaelic) Little fair-haired soldier. #8

Finn (Gaelic) Fair. #7

Finnegan (Gaelic) Fair. #7

Fiske (Old English) Fish. #5

Fitz (Old English) Son of. #7

Fitzgerald (Old English) Son of the spear-mighty. #9

Fitzhugh (Old English) Son of the intelligent man. #6

Fitzpatrick (Old English) Son of a nobleman. #4

Flavian (Latin) Yellow-haired. #2

Fleming (Old English) Dutchman. #3

Fletcher (Old English) Arrow-featherer. #5

Flinn (Gaelic) Son of the red-haired man. #1

Flint (Old English) Stream. #7

Florian (Latin) Flowering or blooming. #3

Floyd (Old Welsh) Grey-haired. #8

Flurry (Gaelic) Leader, prince. #1

Flynn (Gaelic) Son of the red-haired man. #8

Fonda (22/4) (Latin) The profound. #4

Fonsie (Old German) Noble and eager. #5

Fonso (Old German) Noble and eager. #6

Fontaine (Old French) Fountain, spring. #3

Forbes (Gaelic) Prosperous. #2

Ford (Old English) River crossing. #7

Forest (Old French) Forest woodsman. #2

Forrest (Old French) Forest woodsman. #2

Fortino (Latin) Fortunate. #7

Fortunato (Latin) Fortunate. #4

Foster (Latin) Keeper of the woods. #2

Fowler (Old English) Trapper of wild fowl. #7

Francisco (Latin) Free. #7

Frank (Latin) Free man. #5

Franklin (Latin) Free. #4

Franklyn (Latin) Free. #2

Franz (Latin) Free. #2

Fraser (Old English) Curly-haired. #4

Frazer (Old English) Curly-haired. #2

Fred (Old German) Peaceful ruler. #6

Frederic (Old German) Peaceful ruler. #5

Frederick (Old German) Peaceful ruler. #7

Fredric (Old German) Peaceful ruler. #9

Fredrick (Old German) Peaceful ruler. #2

Freeman (Old English) Free. #8

Freemont (Old German) Guardian of freedom. #6

Fritz (Old German) Peaceful ruler. #7

Fulbright (Old German) Very bright. #4

Fuller (Old English) One who shrinks and thickens cloth. #2

Fulton (Old English) A field near the town. #7

Gabe (Hebrew) Devoted to God. #6

Gabian (Hebrew) Devoted to God. #7

Gabor (Hebrew) Devoted to God. #7

Gaelan (22/4) (Gaelic) Intelligent. #4

Gage (old French) Pledge. #2

Galen (Greek) Calm. #3

Gallagher (Gaelic) Eager helper. #8

Galloway (33/6) (Celtic) Stranger from the Gaels. #6

Galvin (Gaelic) Sparrow. #2

Gamal (Arabic) Camel. #7

Gamaliel (33/6) (Hebrew) God gives due reward. #6

Gannon (Gaelic) Fair-complected. #2

Gardner (Old English) A gardener. #4

Gareth (Welsh) Gentle. #5

Garey (Old English) Spear-carrier. #2

Garfield (Old English) Triangular field. #8

Garner (Old French) Armed sentry. #9

Garrad (Old English) Spear-strong. #4

Garrard (Old English) Spear-strong. #4

Garrett (Old German) Spear-mighty. #8

Garrick (Old English) Oak spear. #4

Garrison (Old English) Spear-fortified town. #2

Garron (Old German) Guards, guardian. #1

Garry (33/6) (Old English) Spear-carrier. #6

Garson (Old English) Spear-fortified town. #2

Garth (Old Norse) Groundskeeper. #9

Garvey (33/6) (Gaelic) Rough peace. #6

Garvin (Old German) Spear-friend. #8

Garvyn (33/6) (Old German) Spear-friend ally. #6

Garwood (Old English) From the fir forest. #2

Gary (Old English) Spear-carrier. #6

Gaspar (Persian) Treasurer. #8

Gaston (22/4) (French) From Gascony. #4

Gates (Old English) Gatekeeper. #7

Gavan (Old German) Spear-friend. #9

Gaven (22/4) (Old Welsh) White hawk. #4

Gavin (Old Welsh) White hawk. #8

Gavyn (Old Welsh) White hawk. #6

Gawain (Celtic) The battle hawk. #1

Gayelord (Old French) Lively. #6

Gaylor (33/6) (Old French) Lively. #6

Gaylord (Old French) Lively. #1

Gaynor (Gaelic) Son of the fair-haired man. #8

Genaro (33/6) (Spanish) Consecrated to God. #6

Geno (Hebrew) God's gracious gift. #5

Gentry (Old French) Wellborn. #8

Geoff (Old German) God's divine peace. #3

Geoffrey (Old German) God's divine peace. #6

Geoffry (Old German) God's divine peace. #1

Geomar (Old German) Famous in battle. #5

Geordan (Hebrew) Descending. #1

Geordie (Greek) Landholder, farmer. #9

Georg (Greek) Landholder, farmer. #7

George (Greek) Landholder, farmer. #3

Geovanni (Hebrew) God's gracious gift. #6

Geraint (Celtic) Old. #2

Gerald (Old German) Spear-mighty. #2

Gerard (Old English) Spear-strong. #8

Gerardo (Old English) Spear-strong. #5

Gerhard (Old English) Spear-strong. #7

German (Old German) People of the spear. #4

Geronimo (Latin) Sacred or holy name. #6

Gerrard (Old English) Spear-strong. #8

Gershom (Hebrew) Exile. #4

Gerson (33/6) (Hebrew) Alien resident. #6

Gervais (Old German) Honorable. #9

Gervaise (Old German) Keen with a spear. #5

Gervase (Old German) Honorable. #5

Giacomo (Hebrew) The supplanter. #9

Gian (22/4) (Hebrew) God's gracious gift. #4

Gianni (Hebrew) God's gracious gift. #9

Gibson (Old English) Son of Gilbert. #3

Gideon (Hebrew) Feller of trees. #9

Giff (Old English) Bold giver. #1

Gifford (Old English) Bold giver. #2

Gil (Old German) Brilliant hostage. #1

Gilbert (Old German) Brilliant hostage. #1

Gilberto (Old German) Brilliant hostage. #7

Gilchrist (Gaelic) Servant of Christ. #6

Giles (Greek) Shield-bearer. #7

Gill (22/4) (Old German) Brilliant hostage. #4

Gilmore (Gaelic) Devoted to the Virgin Mary. #7

Gilroy (Gaelic) Devoted to the king. #5

Gino (Hebrew) God's gracious gift. #9

Giomar (Old German) Famous in battle. #9

Giovanni (Hebrew) God is gracious. #1

Girvin (Gaelic) Little rough one. #7

Gitano (Old English) Wanderer. #3

Giuliano (Greek) Young in heart and mind. #7

Giuseppe (Hebrew) He shall add. #8

Gladwin (Old English) Cheerful. #7

Glendon (Scottish Gaelic) From the valley fortress. #8

Glyn (22/4) (Old Welsh) Dweller in a valley. #4

Goddard (Old German) Divinely firm. #8

Godfree (Old German) God's divine peace. #6

Godfrey (Old German) God's divine peace. #8

Godwin (Old English) Friend of God. #9

Gonsales (Old German) Complete warrior. #2

Gonsalve (Old German) Complete warrior. #5

Gonzalo (Old German) Complete warrior. #9

Goodwin (Old English) Friend of God. #6

Gordan (Old English) Round hill. #5

Gorden (Old English) Round hill. #9

Gordie (Old English) Round Hill. #4

Gordon (Old English) Round hill. #1

Gordy (33/6) (Old English) Round hill. #6

Gothart (Old German) Divinely firm. #8

Gowan (Gaelic) Wellborn young warrior. #6

Gradey (33/6) (Gaelic) Noble, industrious. #6

Grady (Gaelic) Noble, illustrious. #1

Graham (Old English) The grey home. #3

Grahame (Old English) The grey home. #8

Gram (Old English) The grey home. #3

Grange (Old English) Farmer. #7

Granger (Old English) Farmer. #7

Grant (French) Great. #6

Grantland (Old English) From the great plains. #1

Grantley (Old English) From the great plains. #3

Granville (Old French) From the large town. #1

Gray (Old English) Son of the bailiff. #6

Grayson (Old English) Son of the bailiff. #9

Greg (Latin) Watchman. #1

Gregg (Latin) Watchman. #8

Gregor (Latin) Watchman. #7

Gregorio (Latin) Watchman. #4

Gregory (Latin) Watchman. #5

Grenville (Old French) From the large town. #5

Grey (Old English) Son of the bailiff. #1

Griff (Old Welsh) Fierce chief. #1

Griffin (Old Welsh) Fierce chief. #6

Griffith (Old Welsh) Fierce chief. #2

Griswold (Old German) From the grey forest. #8

Grover (Old English) From the grove. #4

Guido (Latin) Life. #2

Guillermo (Old German) Determined guardian. #4

Gunnar (German) Battler, warrior. #3

Gunner (German) Battler, warrior. #7

Gunther (German) Battler, warrior. #3

Gus (11/2) (Swedish) Staff of the Goths. #2

Gustave (Swedish) Staff of the Goths. #5

Gustavo (Swedish) Staff of the Goths. #6

Gustavus (22/4) (Swedish) Staff of the Goths. #4

Guthrie (Gaelic) From the windy place. #7

Guy (Old German) Warrior. #8

Gwayne (Celtic) The battle hawk. #3

Gyles (Greek) Shield-bearer. #5

Hadrien (Latin) From Adria. #5

Hadwin (Old English) Battle companion. #5

Hakeem (Arabic) Wise. #7

Hakim (Arabic) Wise. #6

Hal (Old German) Lord of the manor. #3

Hale (Old English) Hero. #8

Halim (Arabic) Gentle. #7

Hall (Old English) From the manor or hall. #6

Halsey (Old English) From Hal's island. #7

Halstead (Old English) From the manor. #7

Hamar (Old Norse) Symbol of ingenuity. #5

Hamid (Arabic) The praised one. #8

Hamilton (Old English) From the proud estate. #2

Hamish (Hebrew) The supplanter. #4

Hamlet (Old French-German) Little home. #5

Hamlin (Old French) The home lover. #3

Hammond (Old German) Small home lover. #5

Hank (Old German) Lord of the manor. #7

Hanley (Old English) From the high pasture. #2

Hans (Hebrew) God's gracious gift. #6

Hansel (Hebrew) God's gracious gift. #5

Harcourt (French) Fortified dwelling. #5

Hardy (Old German) Bold and daring. #2

Hari (Hindi) Princely. #9

Harlan (Old English) From the army land. #9

Harlow (Old English) From the rough hill. #5

Harman (Latin) High-ranking person. #1

Harmon (33/6) (Latin) High-ranking person. #6

Harold (Old German) Lord of the manor. #4

Harper (Old English) Harp player. #3

Harris (Old English) Son of Harry. #1

Harrison (Old English) Son of Harry. #3

Harry (Old German) Lord of the manor. #7

Hart (Old English) Male deer. #2

Harte (Old English) Male deer. #7

Hartley (Old English) From the deer meadow. #8

Harvey (Old German) Army warrior. #7

Hashim (Arabic) Destroyer of evil. #4

Haskel (Hebrew) Understanding. #2

Haslett (22/4) (Old English) From the hazel tree land. #4

Hassan (Arabic) Good-looking. #8

Hastings (Old English) Son of the stern man. #7

Havelock (Old Norse) Sea battle. #5

Hawk (English) A hawk, falcon. #7

Hayden (Old English) From the hedged valley. #3

Hayes (22/4) (Old English) From the hedged place. #4

Hayward (Old English) From the hedged forest. #8

Haywood (Old English) From the hedged forest. #1

Heath (Old English) Heath or wasteland. #6

Heathcliff (Old English) From the heath cliff. #6

Heber (Hebrew) Partner. #2

Hector (33/6) (Greek) Steadfast. #6

Heinrick (Old German) Lord of the manor. #5

Henderson (Old English) Son of Henry. #3

Henri (Old German) Lord of the manor. #9

Henry (Old German) Lord of the manor. #7

Herald (Old English) Army commander. #3

Herbert (Old German) Glorious soldier. #4

Hercules (Greek) Glorious gift. #1

Hereward (Old English) Army guard. #1

Herman (Latin) High-ranking person. #5

Hernando (Old German) World-daring. #7

Herschel (Hebrew) Deer. #6
Hersh (Hebrew) Deer. #4
Hershel (Hebrew) Deer. #3
Hewett (Old French-German) Little and intelligent. #9
Heywood (Old English) From the hedged forest. #5
Hezekiah (Hebrew) God is my strength. #1
Hilario (Latin) Cheerful. #9
Hillard (Old German) Brave warrior. #1
Hillel (Hebrew) Greatly praised. #4
Hilton (33/6) (Old English) From the hill town. #6
Hiram (Hebrew) Most noble. #4
Hirsch (Hebrew) Deer. #2
Hobart (Old German) Bright mind. #1
Hogan (Gaelic) Youthful. #9
Holbrook (Old English) From the brook in the hollow. #6
Holden (Old English) From the hollow in the valley. #4
Holmes (Old English) From the river islands. #9
Holt (Old English) From the forest. #1
Holton (Old English) From the town in the hollow. #3
Homer (Greek) Promise. #5
Horace (Latin) Keeper of the hours. #5
Horatio (Latin) Keeper of the hours. #5
Horst (Old German) A thicket. #8
Horten (Old English) from the grey estate. #8
Horton (Old English) From the grey estate. #9
Hosea (Hebrew) Salvation. #3
Houston (Old English) From the town on the hill. #4
Howard (33/6) (Old English) Cliff guardian. #6
Howe (Old German) High. #6
Howell (Celtic) Little, alert. #3
Howey (Old English) Cliff guardian. #4
Howie (33/6) (Old English) Cliff guardian. #6
Howland (Old English) From the hills. #5
Hubert (Old German) Brilliant, shining mind. #2
Hudson (Old English) Son of the hoodsman. #9
Hugh (Old English) Intelligence. #8
Hughes (Old English) Intelligence. #5
Hugo (Old English) Intelligence. #6
Humbert (33/6) (Old German) Brilliant Hun. #6
Humberto (Old German) Little Hun. #3

Hume (Old English) Home. #2

Humphrey (Old German) Peaceful Hun. #6

Humphry (Old German) Peaceful Hun. #1

Hunt (Old English) A hunter. #9

Hunter (Old English) A hunter. #5

Huntington (Old English) Hunting estate. #7

Huntley (33/6) (Old English) Hunter's meadow. #6

Hurley (Gaelic) Sea tide. #8

Hussein (Arabic) Little and handsome. #5

Hutton (Old English) From the house on the ledge. #8

Huw (Old German) Brilliant thinker. #7

Huxley (Old English) From Hugh's meadow. #5

Hyman (Hebrew) Life. #7

Hymie (33/6) (Hebrew) Life. #6

Hywel (Celtic) Little alert one. #1

Iago (Hebrew) The supplanter. #5

Iaian (Hebrew) God's gracious gift. #7

Iain (Hebrew) God's gracious gift. #6

Ian (Hebrew) God's gracious gift. #6

Ibrahim (Hebrew) Father of the multitude. #6

Ichabod (33/6) (Hebrew) The glory has departed. #6

Ignacio (Latin) Fiery or ardent. #4

Ignatius (Latin) Fiery or ardent. #1

Igor (Norse) Famous son. #4

Ilya ((Hebrew) Jehovah is God. #2

Immanuel (Hebrew) God is with us. #7

Ingemar (Norse) Famous son. #4

Inglebert (Old German) Bright as an angel. #2

Ingram (Old Norse) The son's raven. #8

Innis (Gaelic) From the island. #2

Ira (Hebrew) Watchful. #1

Irvin (Old English) Sea friend. #9

Irving (Old English) Sea friend. #7

Irwin (Old English) Sea friend. #1

Isaac (Hebrew) He laughs. #6

Isaak (Hebrew) He laughs. #5

Isadore (Greek) Gift of Isis. #8

Isaiah (Hebrew) God is my helper. #2

Ishmael (Hebrew) God will hear. #4

Isidor (Greek) Gift of Isis. #2

Isidore (Greek) Gift of Isis. #7

Ismael (Hebrew) God will hear. #5

Israel (Hebrew) Ruling with the Lord. #1

Ivan (Hebrew) God's gracious gift. #1

Ivanhoe (Hebrew) God's gracious gift. #2

Ivar (Old Norse) Battle archer. #5

Ives (Old Norse) Battle archer. #1

Ivon (Old French) Archer with a yew bow. #6

Ivor (Old Norse) Battle archer. #1

Izaak (Hebrew) He laughs. #3

Jabari (Swahili) Valiant. #5

Jabin (Hebrew) God has built. #9

Jacan (11/2) (Hebrew) Trouble. #2

Jacinto (Greek) Hyacinth flower. #9

Jack (Hebrew) The supplanter. #7

Jackson (Old English) Son of Jack. #1

Jacob (Hebrew) The supplanter. #4

Jacques (22/4) (Hebrew) The supplanter. #4

Jadon (Hebrew) Jehovah has heard. #8

Jaedon (22/4) (Hebrew) Jehovah has heard. #4

Jaegar (Old German) Hunter. #6

Jael (Hebrew) Mountain goat. #1

Jaime (Hebrew) The supplanter. #2

Jaimie (Hebrew) The supplanter. #2

Jake (Hebrew) God's gracious gift. #9

Jakeem (Arabic) Raised up. #9

Jamaal (11/2) (Arabic) Beauty. #2

Jamal (Arabic) Beauty. #1

James (Hebrew) The supplanter. #3

Jamil (Arabic) Handsome. #9

Jarad (Hebrew) One who rules. #7

Jareb (Hebrew) He will contend. #9

Jared (Hebrew) One who rules. #2

Jarman (Old German) The German. #3

Jarrett (Old German) Spear-mighty. #2

Jarvis (Old German) Keen with a spear. #7

Jason (Greek) Healer. #5

Jasper (Persian) Treasurer. #6

Javier (Arabic) Bright. #2

Jay (Old French) Blue jay. #9

Jayson (Greek) Healer. #3

Jed (Hebrew) Beloved of the Lord. #1

Jedediah (Hebrew) Beloved of the Lord. #1

Jediah (Hebrew) Jehovah knows. #1

Jedidiah (Hebrew) Beloved of the Lord. #5

Jedreck (Polish) A strong man. #2

Jedrick (33/6) (Polish) A strong man. #6

Jeff (Old German) God's divine peace. #9

Jefferson (Old English) Son of Jeffery. #8

Jeffery (Old German) God's divine peace. #3

Jeffrey (Old German) God's divine peace. #3

Jerald (Old German) Spear-mighty. #5

Jeraldo (Old German) Spear-mighty. #2

Jerard (Old English) Spear-strong. #2

Jerardo (Old English) Spear-strong. #8

Jeremiah (Hebrew) Exalted by the Lord. #6

Jeremias (Hebrew) Exalted by the Lord. #8

Jeremy (Hebrew) Exalted by the Lord. #4

Jeriah (33/6) (Hebrew) Jehovah has seen. #6

Jericho (Arabic) City of the moon. #5

Jermaine (Latin) A German. #3

Jerome (Latin) Sacred or holy name. #3

Jeronimo (Latin) Sacred or holy name. #9

Jerrold (Old German) Spear-mighty. #1

Jervis (Old German) Keen with a spear. #2

Jesus (11/2) (Hebrew) God will help. #2

Jethro (Hebrew) Preeminence. #4

Jim (Hebrew) The supplanter. #5

Jimmy (Hebrew) The supplanter. #7

Jiro (Japanese) Second son. #7

Joab (Hebrew) God is father. #1

Joachim (Hebrew) The Lord will judge. #5

Joaquin (33/6) (Hebrew) Jehovah has established. #6

Job (Hebrew) The afflicted. #9

Joben (Japanese) Enjoy cleanliness. #1

Jock (Hebrew) The supplanter. #3

Joe (Hebrew) He shall add. #3

Joed (Hebrew) Jehovah is witness. #7

Joel (Hebrew) Jehovah is God. #6

John (Hebrew) God's gracious gift. #2

Jomei (Japanese) Spread life. #7

Jon (Hebrew) God's gracious gift. #3

Jonah (Hebrew) Dove. #3

Jonas (Hebrew) Dove. #5

Jonathan (Hebrew) Jehovah gave. #2

Jorge (Greek) Landholder, farmer. #1

Jorgen (33/6) (Greek) Landholder, farmer. #6

Jose (Hebrew) He shall add. #4

Josef (Hebrew) He shall add. #1

Joseph (Hebrew) He shall add. #1

Josephus (Hebrew) He shall add. #5

Josh (Hebrew) Jehovah saves. #7

Joshua (Hebrew) Jehovah saves. #2

Josiah (Hebrew) Jehovah supports. #8

Jotham (22/4) (Hebrew) God is perfect. #4

Jovan (Latin) Father of the sky. #8

Jove (Latin) Father of the sky. #7

Juan (Hebrew) God's gracious gift. #1

Jubal (Hebrew) The ram. #1

Judah (Hebrew) Praised. #8

Judd (Hebrew) Praised. #3

Jude (Latin) Right in the law. #4

Juel (Greek) Youthful, downy-bearded. #3

Jules (Greek) Youthful, downy-bearded. #4

Julian (22/4) (Latin) Related to Julius. #4

Julien (Latin) Related to Julius. #8

Julio (22/4) (Greek) Youthful, downy-bearded. #4

Julius (Greek) Youthful, downy-bearded. #2

Junius (22/4) (Latin) Born in June. #4

Juro (Japanese) Tenth son. #1

Justin (Latin) Just. #3

Justino (Latin) Just. #9

Justis (Latin) Just. #8

Kade (Scottish Gaelic) From the wetlands. #3

Kadir (Arabic) Spring greening. #7

Kaelan (Gaelic) Powerful in battle. #8

Kaemon (Japanese) Joyful. #5

Kaleb (Hebrew) Spontaneous. #4

Kallum (Gaelic) Mild, gentle. #7

Kalvin (Latin) Bald. #6

Kamden (Scottish Gaelic) From the winding valley. #3

Kameron (Scottish Gaelic) Crooked nose. #5

Kane (Gaelic) Tribute. #4

Kanoa (Hawaiian) The free one. #6

Kaori (Japanese) Add a man's strength. #9

Kareem (Arabic) Noble, exalted. #8

Karim (Arabic) Noble, exalted. #7

Karl (Old German) Strong and masculine. #6

Karsten (Greek) Anointed. #7

Kasimir (Slavic) Enforces peace. #8

Kaspar (Persian) Treasurer. #3

Kassim (Arabic) Divided. #9

Keane (Old English) Quick, brave. #9

Kearn (22/4) (Gaelic) Dark. #4

Kearney (Gaelic) Victorious. #7

Keaton (Old English) Place of the hawks. #3

Kedrick (Old English) Gift of splendor. #7

Keefe (Gaelic) Cherished, handsome. #5

Keefer (Gaelic) Cherished, handsome. #5

Keegan (Gaelic) Little and fiery. #7

Keenan (Gaelic) Little and ancient. #5

Keiji (Japanese) Director. #8

Keir (Celtic) Dark-skinned. #7

Keitaro (Japanese) Blessed. #7

Keith (Celtic) From the forest. #8

Kelby (Old German) From the farm by the spring. #1

Kelton (Old English) Town where ships are built. #5

Kelvin (Gaelic) From the narrow river. #1

Ken (Gaelic) Handsome. #3

Kendrick (Old English) Royal ruler. #3

Keneth (Gaelic) Handsome. #9

Kenith (Gaelic) Handsome. #4

Kenji (22/4) (Japanese) Intelligent second son. #4

Kenley (Old English) From the royal meadow. #9

Kenn (Gaelic) Handsome. #8

Kennard (Gaelic) Brave chieftain. #4

Kennedy (33/6) (Gaelic) Helmeted chief. #6

Kenneth (Gaelic) Handsome. #5

Kenny (Gaelic) Handsome. #6

Kenrick (Old English) Bold ruler. #8

Kent (Old Welsh) White, bright. #5

Kentaro (Japanese) Big boy. #3

Kenton (Old English) From the king's estate. #7

Kenyon (Gaelic) White-haired blonde. #3

Keon (Hebrew) God's gracious gift. #9

Keoni (Hebrew) God's gracious gift. #9

Kermit (Gaelic) Free man. #4

Kerr (Old Norse) Marshland. #7

Kerrick (Old English) King's rule. #3

Kerwin (Gaelic) Little, jet-black one. #8

Keven (Gaelic) Gentle, kind, and lovable. #3

Kevin (Gaelic) Gentle, kind, and lovable. #7

Kevyn (Gaelic) Gentle, kind, and lovable. #5

Keye (Gaelic) Son of the fiery one. #1

Khalil (Arabic) Friend. #8

Kian (Hebrew) God's gracious gift. #8

Kiefer (Gaelic) Cherished, handsome. #9

Kiel (Gaelic) Handsome. #1

Kieran (Gaelic) Little and dark-skinned. #4

Kiernan (Gaelic) Little and dark-skinned. #9

Killian (Gaelic) Little and warlike. #5

Kimball (Old Welsh) Warrior chief. #6

Kincaid (33/6) (Celtic) Battle chief. #6

King (Old English) King. #5

Kingsley (Old English) From the king's meadow. #3

Kingston (Old English) From the king's estate. #1

Kirk (22/4) (Old Norse) From the church. #4

Klaus (Greek) Leader of the people. #1

Klinton (Old English) From the headland farm. #5

Knight (33/6) (Old English) Mounted soldier. #6

Knox (Old English) From the hills. #1

Knute (Old Norse) Knot. #8

Koby (Hebrew) The supplanter. #8

Koi (Hawaiian) Urge, implore. #8

Kolby (Old English) From the dark farm. #2

Kolton (Old English) From the coal town. #6

Konnor (33/6) (Gaelic) Wise man. #6

Konrad (Old German) Bold counselor. #9

Korbin (33/6) (Latin) Raven. #6

Kord (Old French) Ropemaker. #3

Kordell (Old French) Ropemaker. #5

Korrigan (Celtic) Spearman. #3

Kort (Old German) Bold. #1

Kraig (Scottish Gaelic) From near the crag. #1

Krishna (Hindi) Delightful. #8

Kristopher (Greek) Bearer of Christ. #4

Kurt (Old German) Bold counselor. #7

Kurtis (Old French) Courteous. #8

Kwasi (African) Born on Sunday. #9

Kyle (Gaelic) Handsome. #8

Kynan (Celtic) High and mighty. #2

Kyne (Old English) The royal one. #1

Lachlan (Scottish Gaelic) From the land of the lakes. #6

Ladd (English) Attendant. #1

Laird (Scottish) Landed proprietor. #8

Lamar (Old German) Famous throughout the land. #9

Lambert (Old German) Bright land. #8

Lamont (Old Norse) Lawyer. #3

Lance (Old French) Servant, attendant. #8

Lancelot (Old French) Servant, attendant. #1

Landan (Old English) From the long hill. #1

Lander (Old English) Owner of the grassland. #9

Landon (Old English) From the long hill. #6

Langdon (Old English) From the long hill. #4

Langley (Old English) From the long meadow. #4

Langston (Old English) From the long, narrow town. #3

Larry (Latin) Crowned with laurels. #2

Lars (Latin) Crowned with laurels. #5

Laszlo (22/4) (Hungarian) Famous leader. #4

Latham (Old Norse) From the barn. #1

Lathrop (Old English) From the barn farmstead. #9

Latimer (33/6) (Old English) Interpreter. #6

Laurence (Latin) Crowned with laurels. #7

Laurens (Latin) Crowned with laurels. #9

Lawford (Old English) From the ford on the hill. #7

Lawrence (Latin) Crowned with laurels. #9

Lawton (22/4) (Old English) From the estate on the hill. #4

Lazarus (Hebrew) God will help. #8

Leander (Greek) Like a lion. #5

Leif (Old Norse) Beloved. #5

Leighton (Old English) From the meadow farm. #9

Leith (Celtic) Broad wide river. #9

Leland (Old English) Meadow land. #3

Lemar (22/4) ((French) Of the sea. #4

Lemuel (Hebrew) Consecrated to God. #5

Lennie (Old French) Lion-brave. #5

Lennon (Gaelic) Little cloak. #2

Lenny (Old French) Lion-brave. #7

Leo (Latin) Lion. #5

Leon (Latin) Lion. #1

Leonard (33/6) (Old French) Lion-brave. #6

Leonardo (Old French) Lion-brave. #3

Leopold (Old German) Bold for the people. #7

Leroy (Old French) King. #3

Les (Latin) From the chosen camp. #9

Lester (Latin) From the chosen camp. #7

Lev (Latin) Lion. #3 .

Levi (Hebrew) Joined, united. #3

Lewis (Old German) Famous warrior. #5

Liam (Old German) Determined guardian. #8

Lincoln (Old English) From the settlement by the pool. #7

Lindon (Old English) From the linden tree. #5

Linford (Old English) From the lime tree ford. #6

Link (Old English) From the settlement by the pool. #1

Linus (Greek) Flaxen-haired. #3

Lionel (Old French) Young lion. #4

Livingston (Old English) From Leif's town. #6

Llewellyn (Welsh) Lionlike or lightning. #3

Lloyd (Old Welsh) Grey-haired. #5

Locke (Old English) From the stronghold or enclosure. #1

Lockwood (Old English) From the enclosed forest. #8

Logan (22/4) (Scottish Gaelic) Little hollow. #4

Lombard (Latin) Long-bearded. #2

Lon (Old German) Noble and eager. #5

London (Old English) Fortress of the moon. #2

Lorenzo (Latin) Crowned with laurels. #6

Lorimer (Old English) Saddlemaker. #9

Loring (Old German) Son of the famous warrior. #3

Lorne (Latin) Crowned with laurels. #1

Lothar (Old German) Famous warrior. #2

Louie (Old German) Famous warrior. #8

Louis (22/4) (Old German) Famous warrior. #4

Lowell (Old French) Little wolf. #7

Lucas (11/2) (Latin) Light. #2

Lucian (Latin) Crowned with laurels. #6

Lucius (22/4) (Latin) Light. #4

Ludlow (Old English) From the prince's hill. #6

Ludovic (Old German) Famous warrior. #5

Ludwig (Old German) Famous warrior. #4

Luis (Old German) Famous warrior. #7

Luke (Latin) Light. #4

Lundy (22/4) (Old French) Born on Monday. #4

Luther (Old German) Famous warrior. #3

Lyle (Old French) From the island. #9

Lyman (Old English) A man from the meadow. #2

Lyndon (Old English) From the linden tree hill. #3

Lyonel (Old French) Young lion. #2

Mac (Scottish Gaelic) Son of. #8

Mace (Old French) Stoneworker. #4

Macey (Old French) Stoneworker. #2

Macnair (Scottish Gaelic) Son of the heir. #5

Maddox (Welsh) Beneficent. #7

Madoc (Welsh) Fortunate. #9

Magnus (Latin) Great. #3

Major (Latin) Greater. #3

Makani (22/4) (Hawaiian) Wind. #4

Makoto (Japanese) Good. #3

Malachi (Hebrew) Angel. #2

Malachy (Greek) Strong and manly. #9

Malcolm (Scottish Gaelic) Disciple of Saint Columbia. #6

Malek (Arabic) Master. #6

Malik (Arabic) Master. #1

Malin (22/4) (Old English) Little warrior. #4

Malvin (Gaelic) Polished chief. #8

Mandel (22/4) (German) Almond. #4

Manfred (Old English) Man of peace. #7

Mano (Hawaiian) Shark. #7

Manolo (Hebrew) God is with us. #7

Manuel (Hebrew) God is with us. #3

Manzo (Japanese) Third son. #6

Marc (Latin) Follower of Mars. #8

Marcel (Latin) Follower of Mars. #7

Marcellus (Latin) Follower of Mars. #5

Marco (Latin) Follower of Mars. #5

Marcus (Latin) Follower of Mars. #3

Marden (Old English) From the valley with the pool. #1

Mareo (Japanese) Rare, uncommon. #7

Mariano (Hebrew) Bitter, graceful. #8

Marino (Latin) From the sea. #7

Mario (Latin) Follower of Mars. #2

Marius (Latin) Follower of Mars. #9

Mark (Latin) Follower of Mars. #7

Marley (Old English) From the hill by the lake. #2

Marmaduke (33/6) (Celtic) Sea leader. #6

Maro (Japanese) Myself. #2

Marsh (Old French) Steward, horse-keeper. #5

Marshal (Old French) Steward, horse-keeper. #9

Marshall (Old French) Steward, horse-keeper. #3

Marsten (Old English) From the town near the marsh. #9

Marston (Old English) From the town near the marsh. #1

Martel (Old French) Mace. #6

Martell (Old French) Mace. #9

Martin (Latin) Follower of Mars. #3

Martyn (Latin) Follower of Mars. #1

Marvin (Old English) Lover of the sea. #5

Mason (Old French)
Stoneworker. #8

Mateo (Hebrew) Gift of God.
#9

Mathew (Hebrew) Gift of God.
#7

Mathias (Hebrew) Gift of God.
#8

Matt (Hebrew) Gift of God. #9

Matthew (Hebrew) Gift of God.
#9

Maurice (Latin) Dark-skinned.
#7

Mauricio (Latin) Dark-skinned.
#8

Max (11/2) (Latin) Most
excellent. #2

Maximilian (Latin) Most
excellent. #6

Maxwell (Old English) Large
well or spring. #9

Mayer (Hebrew) Bringer of
light. #8

Maynard (Old German)
Powerful, brave. #4

Mayo (Gaelic) From the plain
of the yew trees. #9

Mayor (Latin) Greater. #9

McDonald (Celtic) Son of
Donald. #3

Medwin (Old German) Strong
friend. #5

Meir (Hebrew) Bringer of light.
#9

Melbourne (Old English) From
the mill stream. #6

Melvil (Old French) From the
estate of the hard worker.
#1

Melville (Old French) From the
estate of the hard worker.
#9

Melvin (Gaelic) Brilliant chief.
#3

Melvyn (Gaelic) Brilliant chief.
#1

Mendel (Yiddish) Comforter.
#8

Mercer (Old English) Merchant,
storekeeper. #8

Merlin (Old English) Falcon or
hawk. #8

Merlyn (33/6) (Old English)
Falcon or hawk. #6

Merrick (Old English) Ruler of
the sea. #5

Merrill (Old French) Famous.
#6

Merritt (Old English) Worthy.
#4

Merton (Old English) Town by
the sea. #4

Mervin (Old English) Lover of
the sea. #9

Mervyn (Old English) Lover of
the sea. #7

Merwin (Old English) Lover of
the sea. #1

Meyer (Hebrew) Bringer of light. #3

Michio (Japanese) Man with strength of three thousand. #3

Mick (Hebrew) Like unto the Lord. #9

Miguel (Hebrew) Like unto the Lord. #4

Mike (Hebrew) Like unto the Lord. #2

Miles (22/4) (Latin) Soldier, warrior. #4

Millard (33/6) (Latin) Caretaker of the mill. #6

Miller (33/6) (Latin) Caretaker of the mill. #6

Milo (22/4) (Latin) Miller. #4

Milton (Old English) From the mill town. #2

Miner (Latin) Junior, younger. #5

Minor 33/6 (Latin) Junior, younger. #6

Minoru (Japanese) Bear fruit. #9

Mischa (Hebrew) Like unto the Lord. #8

Mitch (Hebrew) Like unto the Lord. #8

Mitchel (Hebrew) Like unto the Lord. #7

Mitchell (Hebrew) Like unto the Lord. #1

Modesto (Latin) Modest, moderate. #1

Moe (Latin) Dark-skinned. #6

Mohammed (Arabic) Praised. #9

Moises (Hebrew) Drawn out of the water. #8

Monolo (Hebrew) God is with us. #3

Monroe (Gaelic) From the mouth of the Roe River. #8

Montague (33/6) (Latin) From the pointed mountain. #6

Monte (22/4) (Latin) From the pointed mountain. #4

Montgomery (Old English) From the rich man's mountain. #1

Monti (Latin) From the pointed mountain. #8

Monty (Latin) From the pointed mountain. #6

Moore (Old French) Dark-complected. #3

Mordecai (Hebrew) Warrior. #5

Morell (Old French) The Moor. #3

Morey (Latin) Dark-skinned. #4

Morio (Japanese) Forest boy. #7

Morland (Old English) Marsh, wetland. #5

Morley (Old English) From the meadow on the moor. #7

Morly (Old English) From the meadow on the moor. #7

Morris (Latin) Dark-skinned. #2

Morse (Old English) Son of the dark-complexioned one. #7

Mort (Old French) Still water. #3

Mortie (Old French) Still water. #8

Mortimer (Old French) Still water. #3

Morton (Old English) From the town on the moor. #5

Morty (Old French) Still water. #1

Morven (33/6) (Gaelic) Blond giant. #6

Moses (Hebrew) Drawn out of the water. #8

Moshe (Hebrew) Drawn out of the water. #6

Moss (Hebrew) Drawn out of the water. #3

Muhammad (Arabic) Praised. #2

Muir (Scottish Gaelic) Moor. #7

Munro (Gaelic) From the mouth of the Roe River. #9

Munroe (Gaelic) From the mouth of the Roe River. #5

Murdoch (Scottish Gaelic) Sailor. #1

Murdock (Scottish Gaelic) Wealthy sailor. #4

Murray (33/6) (Scottish Gaelic) Sailor. #6

Murry (Scottish Gaelic) Sailor. #5

Myer (Hebrew) Bringer of light. #7

Myles (Latin) Soldier, warrior. #2

Myron (Greek) Fragrant ointment. #4

Nahum (Hebrew) Consoling. #3

Nairn (Celtic) Dweller by the alder tree. #2

Namir (Arabic) Leopard. #1

Napoleon (Greek) Lion of the woodland dell. #2

Nardo (Latin) Strong, hardy. #7

Nat (Hebrew) Gift of God. #8

Natal (Latin) Natal day. #3

Natale (Latin) (Latin) Natal day. #8

Nate (Hebrew) Gift of God. #4

Nathan (22/4) (Hebrew) Gift of God. #4

Nathaniel (Hebrew) Gift of God. #3

Navarro (Old Spanish) Plains. #8

Neal (Gaelic) The champion. #5

Neale (Gaelic) The champion. #1

Neall (Gaelic) The champion. #8

Ned (Old English) Rich and happy protector. #5

Nehemiah (Hebrew) Compassion of Jehovah. #9

Neil (22/4) (Gaelic) The champion. #4

Nels (Gaelic) The champion. #5

Nelson (English) Neil's son. #7

Nemo (Greek) From the glen. #2

Nero (Latin) Stern. #7

Neron (Latin) Stern. #3

Nestor (Greek) Traveler, wisdom. #1

Neuman (Old English) New arrival. #5

Neville (Latin) From the new town. #7

Nevin (Gaelic) Holy, worshiper of the saints. #1

Nevins (Gaelic) Holy, worshiper of the saints. #2

Newbold (Old English) From the new building. #3

Newman (Old English) New arrival. #7

Newton (Old English) From the new town. #1

Niall (Gaelic) The champion. #3

Nicholas (Greek) The people's victory. #9

Nick (Greek) The people's victory. #1

Nicolas (Greek) The people's victory. #1

Niels (Gaelic) The champion. #5

Nigel (Latin) Black-haired. #2

Niles (Gaelic) The champion. #5

Nixon (Greek) The people's victory. #4

Noach (Hebrew) Rest, comfort. #5

Noah (Hebrew) Rest, comfort. #2

Noble (Latin) Wellborn. #3

Noe (Hebrew) Rest, comfort. #7

Nolan (Gaelic) Famous or noble. #2

Noland (Gaelic) Famous or noble. #6

Norbert (Old Norse) Brilliant hero. #2

Norbie (Old Norse) Brilliant hero. #9

Norby (Old Norse) Brilliant hero. #2

Norm (Old French) A northman. #6

Norman (Old French) A northman. #3

Normand (Old French) A northman. #7

Normie (Old French) A northman. #2

Normy (Old French) A northman. #4

Norrie (Old French) Northerner. #7

Norris (Old French) Northerner. #3

Norry (Old French) Northerner. #9

North (Old English) From the north farm. #3

Northrop (Old English) From the north farm. #7

Northrup (Old English) From the north farm. #4

Norton (33/6) (Old English) From the northern town. #6

Norward (Old English) Guardian of the north. #3

Nye (Welsh) Honor or gold. #8

Nyles (Gaelic) The champion. #3

Oakes (Old English) From the oak trees. #6

Oakley (Old English) From the oak tree field. #6

Oates (Greek) Keen of hearing. #6

Obadiah (Hebrew) Servant of God. #4

Obadias (Hebrew) Servant of God. #6

Obed (Hebrew) Servant of God. #8

Obediah (Hebrew) Servant of God. #8

Octave (Latin) The eighth. #3

Octavio (Latin) The eighth. #4

Octavius (Latin) The eighth. #2

Ogdan (Old English) From the oak valley. #5

Ogden (Old English) From the oak valley. #9

Olaf (Old Norse) Ancestral talisman. #7

Olav (Old Norse) Ancestral talisman. #5

Ole (Old Norse) Ancestral talisman. #5

Olin (Old Norse) Ancestral talisman. #5

Oliver (Latin) Symbol of peace. #9

Olivero (Latin) Symbol of peace. #6

Olivier (Latin) Symbol of peace. #9

Oliviero (Latin) Symbol of peace. #6

Ollie (Latin) Symbol of peace. #8

Olly (Latin) Symbol of peace. #1

Olvan (Latin) Symbol of peace. #1

Omar (Arabic) Most high follower of the Prophet. #2

Omarr (Arabic) Most high follower of the Prophet. #2

Oran (Gaelic) Pale-complected. #3

Orel (Russian) Eagle. #5

Oren (Gaelic) Pale-complected. #7

Orestes (Greek) Mountain man. #2

Orin (Gaelic) Pale-complected. #2

Orion (Latin) Dawning, golden. #8

Orlan (Old English) From the pointed land. #6

Orland (Old English) From the pointed land. #1

Orlando (Old English) From the pointed land. #7

Orman (Old German) Mariner or shipman. #7

Ormond (Old English) Spear-protector. #7

Orren (Gaelic) Pale-complected. #7

Orrin (Gaelic) Pale-complected. #2

Orsino (Latin) Bearlike. #9

Orson (Latin) Bearlike. #9

Orv (Old French) From the golden estate. #1

Orval (Old English) Mighty spear. #5

Orville (Old French) From the golden estate. #3

Orvin 33/6 (Old English) Spear-friend. #6

Osborn (Old English) Warrior of God. #2

Osborne (Old English) Warrior of God. #7

Osbourn (Old English) Warrior of God. #5

Osbourne (Old English) Warrior of God. #1

Oscar (Old English) Divine spearman. #2

Osgood (Old English) Divinely good. #3

Oskar (Old English) Divine spearman. #1

Osman (Old English) God's servant. #8

Osmond (Old English) Divine protector. #8

Osmund (Old English) Divine protector. #5

Ossie (22/4) (Old English) Divine spearman. #4

Ossy (Old English) Divine spearman. #6

Oswald (Old English) Divinely powerful. #2

Oswell (Old English) Divinely powerful. #5

Otes (Greek) Keen of hearing. #5

Otho (22/4) (Old German) Rich. #4

Otis (Greek) Keen of hearing. #9

Otto (Old German) Rich. #7

Owain (Old Welsh) Wellborn young warrior. #8

Owen (Old Welsh) Wellborn young warrior. #3

Oxford (Old English) From the river-crossing of the oxen. #1

Ozzie (Old English) Divine spearman. #9

Pablo (Latin) Little. #1

Packston (Latin) From the peaceful town. #9

Paco (Latin) Free. #8

Paddy (Latin) Wellborn, noble. #5

Padraig (Latin) Wellborn, noble. #2

Paget (22/4) (French) Useful assistant. #4

Paine (Latin) Country man. #9

Palladin (33/6) (American Indian) Fighter. #6

Pancho (Latin) Free. #3

Paolo (Latin) Little. #5

Park (Old English) Guardian of the park. #1

Parke (Old English) Guardian of the park. #6

Parker (33/6) (Old English) Guardian of the park. #6

Parkin (33/6) (Old English) Little Peter. #6

Parnel (Old French) Little Peter. #3

Parnell (33/6) (Old French) Little Peter. #6

Parrish (Old English) From the churchyard. #8

Parrnell (Old French) Little Peter. #6

Parry (33/6) (Old English) From the churchyard. #6

Parsifal (Old French) Valley-piercer. #1

Pascal (Latin) Born at Easter or Passover. #7

Pascual (Latin) Born at Easter or Passover. #1

Pasquale (Latin) Born at Easter or Passover. #2

Paton (Latin) Wellborn, noble. #3

Patric (Latin) Wellborn, noble. #4

Patrice (Latin) Wellborn, noble. #9

Patricio (Latin) Wellborn, noble. #1

Patrick (33/6) (Latin) Wellborn, noble. #6

Patton (Old English) From the warrior's estate. #5

Paul (Latin) Little. #5

Paxon (Latin) From the peaceful town. #7

Paxton (Latin) From the peaceful town. #9

Payne (Latin) Villager. #7

Pearce (Latin) Rock or stone. #3

Pedraic (Latin) Wellborn, noble. #2

Pedro (Latin) Rock or stone. #4

Pembroke (Celtic) From the headland. #4

Pembrook (Celtic) From the headland. #5

Penn (22/4) (Old English) Enclosure. #4

Pehrod (Old German) Famous commander. #9

Pepin (33/6) (Old German) Petitioner. #6

Per (Latin) Rock or stone. #3

Perceval (Old French) Valley-piercer. #1

Percival (Old French) Valley-piercer. #5

Percy (Old French) Valley-piercer. #4

Pernell (Old French) Little Peter. #1

Perrie (Old French) Pear tree. #8

Perry (Old French) Pear tree. #1

Pete (Latin) Rock or stone. #1

Peter (Latin) Rock or stone. #1

Peyton (Old English) From the warrior's estate. #5

Phelan (Gaelic) Brave as a wolf. #2

Phil (Greek) Lover of horses. #9

Philip (Greek) Lover of horses. #7

Phillip (Greek) Lover of horses. #1

Philo (33/6) (Greek) Loving, friendly. #6

Phineas (Hebrew) Oracle. #9

Pierce (Latin) Rock or stone. #2

Pierre (Latin) Rock or stone. #8

Pierrot (Latin) Rock or stone. #2

Piers (Latin) Rock or stone. #4

Pincas (Hebrew) Oracle. #8

Pinchas (Hebrew) Oracle. #7

Pincus (Hebrew) Oracle. #1

Placido (33/6) (Latin) Peaceful, serene. #6

Plato (Greek) Broad-shouldered. #1

Porfirio (Greek) Dressed in purple. #7

Porter (Latin) Door keeper. #2

Powell (Celtic) Alert. #2

Prentice (Old English) Apprentice. #9

Prescott (Old English) From the priest's cottage. #8

Preston (Old English) From the priest's estate. #8

Price (33/6) (Old Welsh) Son of the ardent one. #6

Primo (Latin) First, first child. #8

Prince (Latin) Chief, prince. #2

Prinz (Latin) Chief, prince. #2

Procter (Latin) Administrator. #5

Proctor (Latin) Administrator. #6

Pryor (Latin) Head of a monastery. #2

Purcell (33/6) (Old French) Valley-piercer. #6

Purvis (33/6) (English-French) To provide food. #6

Putnam (22/4) (Old English) Dweller by the pond. #4

Quentin (Latin) Fifth child. #1

Quillan (Gaelic) Cub. #5

Quillon (Latin) The sword. #1

Quincy (Old French) From the fifth son's estate. #8

Quinlan (Gaelic) Wise, intelligent. #7

Quint (Latin) Fifth child. #9

Quintin (Latin) Fifth child. #5

Quinton (Latin) Fifth child. #2

Quintus (Latin) Fifth Child. #4

Rad (Old English) From the red stream. #5

Radburn (33/6) (Old English) From the red stream. #6

Radcliffe (Old English) From the red cliff. #1

Raddie (Old English) From the red stream. #5

Raddy (Old English) From the red stream. #7

Radford (Old English) From the reedy ford. #3

Rafael (Hebrew) Healed by God. #7

Rafaello (Hebrew) Healed by God. #7

Rafe (Hebrew) Healed by God. #3

Raff (22/4) (Old English) Swift wolf. #4

Rafferty (Gaelic) Rich and prosperous. #9

Rafi (Arabic) Exalting. #7

Rai (Japanese) Lightning, thunder. #1

Raimondo (Old German) Mighty or wise protector. #8

Rainer (Old German) Mighty army. #2

Rajan (Sanskrit) King. #8

Raleigh (Old English) From the deer meadow. #6

Ralf (Old English) Swift wolf. #1

Ralph (Old English) Swift wolf. #1

Ram (Old English) From the Ram's island. #5

Rami (Arabic) Loving. #5

Ramiro (Spanish) Wise, renowned. #2

Ramon (Old German) Mighty or wise protector. #7

Ramsay (Old English) From the ram's island. #5

Ramsey (Old English) From the ram's island. #9

Rance (African) Borrowed all. #5

Rancell (African) Borrowed all. #2

Rand (Old English) Shield-wolf. #1

Randal (Old English) Shield-wolf. #5

Randall (Old English) Shield-wolf. #8

Randolf (Old English) Shield-wolf. #7

Randolph (Old English) Shield-wolf. #7

Ranger (Old French) Guardian of the forest. #9

Ranier (Old Norse) Mighty Army. #2

Ransell (African) Borrowed all. #9

Ransom (Old English) Son of the shield. #8

Ranulf (Old English) Shield-wolf. #9

Raoul (22/4) (Old English) Swift wolf. #4

Raphael (Hebrew) Healed by God. #7

Rashad (Arabic) Counselor. #6

Rasheed (33/6) (Arabic) Counselor. #6

Rashid (Arabic) Counselor. #5

Raul (Old English) Swift wolf. #7

Ravi (Hindi) Sun. #5

Ravid (Hindi) Sun. #9

Raviv (Hindi) Sun. #9

Rawley (Old English) From the deer meadow. #3

Rawlins (33/6) (Anglo-French) Son of little counsel-wolf. #6

Rayburn (Old English) From the deer brook. #9

Raymond (Old German) Mighty or wise protector. #9

Raymund (33/6) (Old German) Mighty or wise protector. #6

Raynor (Old German) Mighty army. #1

Read (Old English) Red-haired. #1

Reade (Old English) Red-haired. #6

Red (Old English) Red-colored. #9

Redd (22/4) (Old English) Red-colored. #4

Redford (Old English) Red river crossing. #7

Redman (Old German) Protecting counselor. #1

Redmond (Old German) Protecting counselor. #1

Redmund (Old German) Protecting counselor. #7

Reece (Welsh) Enthusiastic. #9

Reed (Old English) Red-haired. #5

Reede (Old English) Red-haired. #1

Reeve (Old English) Steward. #1

Refugio (Spanish) Refuge, shelter. #9

Reginald (Old English) Mighty and powerful. #7

Regis (Latin) Rules. #4

Rei (Japanese) Law, rule. #5

Reid (Old English) Red-haired. #9

Reinhard (Old French) Fox. #5

Remington (Old English) From the raven estate. #7

Remus (22/4) (Latin) Fast-moving. #4

Renard (33/6) (Old French) Fox. #6

Renato (Latin) Reborn. #1

Renjiro (Japanese) Clean, upright, honest. #8

Reuben (Hebrew) Behold a son. #2

Reuven (Hebrew) Behold a son. #4

Rex (Latin) King. #2

Rexford (Old English) King's river crossing. #9

Rey (Latin) King. #3

Reyes (Latin) Kings. #9

Reynaldo (Old English) Mighty and powerful. #4

Reynard (Old French) Fox. #4

Reynato (Old English) Mighty and powerful. #8

Reynold (Old English) Mighty and powerful. #3

Rhett (Welsh) Enthusiastic. #8

Rhodes (33/6) (Old English) Dweller at the crucifixes. #6

Rhys (Welsh) Enthusiastic. #7

Ricardo (Old German) Wealthy, powerful. #5

Riccardo (Old German) Wealthy, powerful. #8

Ricco (Old German) Wealthy, powerful. #3

Rich (Old German) Wealthy, powerful. #2

Richard (Old German) Wealthy, powerful. #7

Richmond (Old German) Powerful protector. #3

Richy (Old German) Wealthy, powerful. #9

Rick (Old German) Wealthy, powerful. #5

Rico (Old German) Wealthy, powerful. #9

Rider (Old English) Horseman. #9

Ridley (Old English) From the red meadow. #1

Rigby (Old English) Valley of the rulers. #7

Rigel (33/6) (Arabic) Foot. #6

Ring (Old English) Ring. #3

Ringo (Old English) Ring. #9

Rinji (33/6) (Japanese) Peaceful forest. #6

Riordan (Gaelic) Bard, royal poet. #7

Rip (Old English) From the shouter's meadow. #7

Ripley (Old English) From the shouter's meadow. #4

Ritchie (Old German) Wealthy, powerful. #9

Roald (Scandinavian) Renowned, powerful. #5

Roan (Old English) From the rowan tree. #3

Roane (Gaelic) Red-haired, red. #8

Roarke (Gaelic) Famous ruler. #5

Rob (Old English) Bright fame. #8

Robb (Old English) Bright fame. #1

Robert (33/6) (Old English) Bright fame. #6

Roberto (Old English) Bright fame. #3

Robertson (English) Son of Robert. #9

Robinson (English) Son of Robin. #7

Rochester (Old English) From the stone camp. #3

Rock (Old English) From the rock. #2

Rockwell (Old English) From the rocky spring. #9

Rod (Old English) One who rides with a knight. #1

Rodd (Old English) One who rides with a knight. #5

Roddy (Old English) One who rides with a knight. #3

Roden (Old English) From the reed valley. #2

Roderic (Old German) Famous ruler. #9

Roderick (Old German) Famous ruler. #2

Roderigo (Old German) Famous ruler. #1

Rodger (Old German) Famous spearman. #4

Rodin (33/6) (Old English) From the reed valley. #6

Rodman (Old English) One who rides with a knight. #2

Rodney (Old English) From the island clearing. #9

Rodolfo (Old German) Famous wolf. #4

Rogan (Gaelic) Red-haired. #1

Rogelio (Old German) Famous spearman. #9

Roger (Old German) Famous spearman. #9

Rogers (Old German) Famous spearman. #1

Rohan (Gaelic) Red-haired. #2

Roka (Japanese) White crest of the wave. #9

Roland (Old German) From the famous land. #1

Rolando (Old German) From the famous land. #7

Rolf (Old English) Swift-wolf. #6

Rolfe (Old English) Swift-wolf. #2

Rollo (Old German) From the famous land. #9

Rolph (33/6) (Old English) Swift-wolf. #6

Romain (Latin) From Rome. #7

Roman (Latin) From Rome. #7

Romeo (Italian) Pilgrim to Rome. #3

Romulus (Latin) Citizen of Rome. #2

Ron (Old English) Mighty and powerful. #2

Ronald (Old English) Mighty and powerful. #1

Ronan (Gaelic) Little seal. #8

Rooney (Gaelic) Red-haired. #2

Roosevelt (Old Dutch) From the rose field. #5

Rory (Gaelic) Red king. #4

Rosco (Old Norse) From the deer forest. #7

Roscoe (Old Norse) From the deer forest. #3

Ross (Celtic) From the peninsula. #8

Roswald (Old English) From a field of roses. #2

Roswell (Old English) From a field of roses. #5

Roth (Old German) Red hair. #7

Rourke (Gaelic) Famous ruler. #7

Rowland (33/6) (Old German) From the famous land. #6

Roy (22/4) (Old French) King. #4

Royal (Old French) Royal. #8

Royall (Old French) Royal. #2

Royce (Old English) Son of the king. #3

Royd (Old Norse) From the forest clearing. #8

Ruben (Hebrew) Behold a son. #6

Rudd (Old English) From the red enclosure. #2

Ruddy (Old English) From the red enclosure. #9

Rudolf (Old German) Famous wolf. #4

Rudolph (Old German) Famous wolf. #4

Rudy (Old German) Famous wolf. #5

Rudyard (Old English) From the red enclosure. #1

Rufino (Latin) Red-haired. #2

Rufus (22/4) (Latin) Red-haired. #4

Rupert (Old English) Bright fame. #8

Ruprecht (Old English) Bright fame. #1

Rurik (Old German) Famous ruler. #5

Rush (Old French) Red-haired. #3

Rushford (Old English) From the rush ford. #1

Ruskin (Old French) Red-haired. #2

Russ (Old French) Red-haired. #5

Russel (22/4) (Old French) Red-haired. #4

Russell (Old French) Red-haired. #7

Rustin (Old French) Red-haired. #2

Rutger (Old German) Famous spearman. #8

Rutherford (Old English) From the cattle ford. #7

Rutland (Old Norse) From the stumpland. #9

Rutledge (Old English) From the red pool. #2

Ruttger (Old German) Famous spearman. #1

Ruy (Old French) King. #1

Ryder (Old English) Horseman. #7

Ryker (Old German) Wealthy, powerful. #5

Ryleigh (Gaelic) Valiant. #3

Sabin (Latin) Sabine man. #9

Sachio (Japanese) Fortunately born. #1

Sage (Old French) Wise one. #5

Saleem (Arabic) Safe, peace. #1

Salim (Arabic) Safe, peace. #9

Salomon (Hebrew) Peaceful. #8

Salvador (Italian) Savior. #2

Salvator (Italian) Savior. #9

Salvatore (Italian) Savior. #5

Salvidor (Italian) Savior. #1

Sam (Hebrew) His name is God. #6

Sampson (Hebrew) Like the sun. #7

Samson (Hebrew) Like the sun. #9

Samuel (Hebrew) His name is God. #8

Sanborn (Old English) From the sandy brook. #2

Sancho (Latin) Sanctified. #6

Sander (Old English) Son of Alexander. #7

Sanders (Old English) Son of Alexander. #8

Sanderson (Old English) Son of Alexander. #1

Sandford (Old English) From the sandy hill. #9

Sanford (Old English) From the sandy hill. #5

Sanjiro (Japanese) Admirable. #5

Sanson (Hebrew) Like the sun. #1

Sansone (Hebrew) Like the sun. #6

Santiago (Spanish) Saint James. #5

Santino (Latin) Saints. #2

Santos (Latin) Saints. #7

Sargent (Old French) Army officer. #3

Saul (Hebrew) Asked for. #8

Sauncho (Latin) Sanctified. #9

Saunders (Old English) Son of Alexander. #2

Sawyer (Old English) Sawer of wood. #1

Saxon (Old English) Swordsman. #1

Sayer (Welsh) Carpenter. #5

Sayers (Welsh) Carpenter. #6

Sayre (Welsh) Carpenter. #5

Sayres (Welsh) Carpenter. #6

Schuyler (Dutch) Sheltering. #3

Scot (Old English) Scotsman. #3

Scott (Old English) Scotsman. #5

Seamus (Hebrew) The supplanter. #6

Sebastian (Latin) Venerated. #9

Seiji (Japanese) Lawful. #7

Selby (Old English) From the village by the mansion. #9

Seldon (Old English) From the willow tree valley. #6

Selig (Old German) Blessed. #7

Selwin (Old English) Friend from the manor house. #1

Selwyn (Old English) Friend from the manor house. #8

Septimus (Latin) Seventh-born son. #5

Serge (Latin) The attendant. #9

Sergei (Latin) The attendant. #9

Sergent (Old French) Army officer. #7

Sergio (Latin) The attendant. #1

Sergius (Latin) The attendant. #8

Serle (Old German) Bearer of arms. #5

Seth (Hebrew) Appointed, substitute. #7

Seton (Old English) From the place by the sea. #1

Severin (Latin) Strict, restrained. #2

Severo (Latin) Strict, restrained. #3

Seward (Old English) Sea guardian. #7

Sewell (22/4) (Old English) Sea powerful. #4

Sexton (Old English) Church official. #7

Sextus (Latin) Sixth-born. #9

Seymour (Old French) From the town of Saint Maur. #8

Shaan (Hebrew) Peaceful. #7

Shalom (Hebrew) Peace. #5

Shamus (Hebrew) The supplanter. #9

Sharif (Arabic) Illustrious. #7

Shaun (Hebrew) God's gracious gift. #9

Shaw (Old English) From the grove. #6

Sheehan (33/6) (Gaelic) Little and peaceful. #6

Sheffield (Old English) From the crooked field. #2

Shelden (Old English) From the town on the ledge. #4

Sheldon (Old English) From the town on the ledge. #5

Shem (Hebrew) His name is God. #9

Shep (Old English) Shepherd. #3

Shepard (Old English) Shepherd. #8

Shepherd (Old English) Shepherd. #2

Sherborn (Old English) From the clear brook. #9

Sherlock (Old English) Fair-haired. #1

Sherman (33/6) (Old English) Wool-shearer. #6

Sherwin (Old English) Swift runner. #6

Sherwood (Old English) From the bright forest. #8

Sid (Old French) From Saint Denis. #5

Siegfried (Old German) Victorious protector. #1

Sig (Old German) Victorious protector. #8

Sigfried (Old German) Victorious protector. #5

Siggy (Old German) Victorious protector. #4

Sigmund (33/6) (Old German) Victorious protector. #6

Silas (Latin) Forest God. #6

Silvano (Latin) From the forest. #2

Silvester (Latin) From the forest. #3

Silvio (Latin) From the forest. #5

Simon (Hebrew) One who hears. #7

Simpson (33/6) (Hebrew) Like the sun. #6

Sinclair (Old French) From Saint Clair. #4

Sinclare (Old French) From Saint Clair. #9

Skeeter (Old English) The swift. #2

Skell (Gaelic) Storyteller. #5

Skelly (Gaelic) Storyteller. #3

Skerry (33/6) (Scandinavian) From the rocky island. #6

Skip (Scandinavian) Shipmaster. #1

Skipper (Scandinavian) Shipmaster. #4

Skippy (33/6) (Scandinavian) Shipmaster. #6

Skipton (Scandinavian) Shipmaster. #5

Slade (Old English) Child of the valley. #5

Slevin (Gaelic) The mountain climber. #9

Sly (11/2) (Latin) From the forest. #2

Smith (Old English) Blacksmith. #6

Smitty (Old English) Blacksmith. #7

Sol (Hebrew) Peaceful. #1

Solly (Hebrew) Peaceful. #2

Solomon (Hebrew) Peaceful. #4

Somerset (33/6) (Old English) Place of the summer settlers. #6

Sonnie (English) Son. #4

Sonny (English) Son. #6

Sorrel (33/6) (Old French) Reddish-brown. #6

Spark (Old English) Happy. #2

Sparke (Old English) Happy. #7

Sparky (Old English) Happy. #9

Spence (Old English) Dispenser of provisions. #8

Spencer (Old English) Dispenser of provisions. #8

Spense (Old English) Dispenser of provisions. #6

Sprague (33/6) (French) Lively. #6

Stafford (Old English) From the riverbank landing. #8

Stamford (33/6) (Old English) From the stormy crossing. #6

Stan (Old English) From the rocky meadow. #9

Stanford (Old English) From the rocky ford. #7

Stanislaus (Slavic) Stand of glory. #9

Stanislaw (Slavic) Stand of glory. #1

Stanleigh (Old English) From the rocky meadow. #5

Stanley (Old English) From the rocky meadow. #6

Stanton (22/4) (Old English) From the rocky farm. #4

Stanway (22/4) (Old English) From the rocky road. #4

Stanwood (Old English) From the rocky wood. #3

Steele (Old English) Hard, durable. #3

Stefan (Greek) Crowned. #2

Stefano (Greek) Crowned. #8

Stein (22/4) (Old German) Rock. #4

Steinar (Old German) Rock warrior. #5

Steiner (Old German) Rock warrior. #9

Stephan (Greek) Crowned. #2

Stephen (33/6) (Greek) Crowned. #6

Sterling (Old English) Valuable. #5

Stern (22/4) (Old English) Austere. #4

Sterne (Old English) Austere. #9

Steve (Greek) Crowned. #8

Steven (22/4) (Greek) Crowned. #4

Steward (Old English) The steward. #9

Stewart (Old English) The steward. #7

Stillman (Old English) Quiet man. #1

Stillmann (33/6) (Old English) Quiet man. #6

Stirling (Old English) Valuable. #9

Stone (Old English) Stone. #1

Stoney (Old English) Stone. #8

Stuart (Old English) The steward. #9

Styles (Old English) From the stiles. #1

Sullivan (Gaelic) Black-eyed. #2

Sully (Gaelic) Black-eyed. #8

Sumner (Old English) Church officer, summoner. #9

Sutherland (Scandinavian) From the southern land. #5

Sutton (Old English) From the southern town. #1

Sven (Scandinavian) Youth. #6

Sweeney (33/6) (Gaelic) Little hero. #6

Swen (Scandinavian) Youth. #7

Swinton (33/6) (Old English) From the pig farm. #6

Sylvan (Latin) From the forest. #3

Sylvester (Latin) From the forest. #1

Tab (Old English) Drummer. #5

Taber (Old English) Drummer. #1

Tabib (Turkish) Physician. #7

Tabor (Old English) Drummer. #2

Tad (Greek) Stout-hearted. #7

Tadashi (Japanese) Serves the master faithfully. #8

Tadd (11/2) (Greek) Stout-hearted. #2

Tadeo (Latin) Praise. #9

Taite (Old English) Cheerful. #1

Taj (Hindi) Crown. #4

Taji (Japanese) Silver and yellow color. #4

Takeo (Japanese) Strong like bamboo. #7

Talbert (Old German-French) Valley, bright. #6

Talbot (Old German-French) Valley, bright. #7

Tally (Old German-French) Valley, bright. #7

Tamas (Hebrew) The devoted brother. #9

Tanjiro (33/6) (Japanese) High-valued second son. #6

Tanner (Old English) Leather worker, tanner. #9

Taro (Japanese) Firstborn male. #9

Tau (African) Lion. #6

Tavis (Hebrew) The devoted brother. #8

Tavish (Gaelic) Twin. #7

Tayib (Indian) Good or delicate. #3

Teague (Gaelic) Bard. #5

Ted (11/2) (Greek) Gift of God. #2

Teiji (Japanese) Righteous second son. #8

Templeton (Old English) From the town of the temple. #3

Teo (Greek) Gift of God. #4

Terance (Latin) Smooth. #3

Terence (Latin) Smooth. #7

Terencio (Latin) Smooth. #8

Terran (English) Earth man. #4

Terrance (Latin) Smooth. #3

Terrell (Old German) Martial. #9

Terrence (Latin) Smooth. #7

Terrill (Old German) Martial. #4

Thaddeus (Greek) Stout-hearted. #1

Thaine (Old English) Attendant warrior. #3

Thane (Old English) Attendant warrior. #3

Thatcher (Old English) Roof thatcher. #2

Thaxter (33/6) (Old English) Roof thatcher. #6

Thayer (Old French) From the nation's army. #5

Thayne (Old English) Attendant warrior. #1

Theo (Greek) Gift of God. #3

Theobald (Old German) Ruler of the people. #4

Theodore (Greek) Gift of God. #9

Theodoric (Old German) Ruler of the people. #7

Theon (Greek) Godly man. #8

Theron (Greek) The hunter. #8

Thom (Hebrew) The devoted brother. #2

Thomas (22/4) (Hebrew) The devoted brother. #4

Thor (Old Norse) Thunder. #7

Thorald (33/6) (Scandinavian) Thor's ruler. #6

Thorn (Old English) From the thorny embankment. #3

Thorndike (Old English) From the thorny embankment. #5

Thorne (Old English) From the thorny embankment. #8

Thornton (Old English) From the thorny farm. #7

Thorpe (Old English) From the village. #1

Thorstein (Scandinavian) Thor's stone. #2

Thurston (Scandinavian) Thor's stone. #9

Tiler (Old English) Maker of tiles. #1

Tim (Greek) Honoring God. #6

Timmie (33/6) (Greek) Honoring God. #6

Timmy (Greek) Honoring God. #8

Timon (Greek) Honoring God. #8

Timothy (Greek) Honoring God. #2

Tito (Greek) Of the giants. #1

Titos (Greek) Of the giants. #2

Titus (Greek) Of the giants. #8

Tobias (Hebrew) The Lord is good. #3

Tod (Old English) A fox. #3

Todd (Old English) A fox. #7

Toddie (Old English) A fox. #3

Toddy (Old English) A fox. #5

Tom (Hebrew) The devoted brother. #3

Tomas (Hebrew) The devoted brother. #5

Tomaso (Hebrew) The devoted brother. #2

Tomeo (Japanese) Cautious man. #5

Tomkin (Hebrew) The devoted brothers. #1

Tomlin (Hebrew) The devoted brother. #2

Tommy (Hebrew) The devoted brother. #5

Tonio (Latin) Praiseworthy, without a peer. #1

Torin (Gaelic) Chief. #4

Torrance (Gaelic) From the knolls. #4

Toru (Japanese) Sea. #2

Toshiro (Japanese) Talented. #5

Townsend (33/6) (Old English) From the town's end. #6

Trahern (Welsh) Strong as iron. #3

Traver (Old French) At the crossroads. #3

Travers (Old French) At the crossroads. #4

Travis (Old French) At the crossroads. #8

Tray (Welsh) Strong as iron. #1

Trefor (Gaelic) Wise and discreet. #1

Tremaine (Celtic) From the house of stone. #4

Tremayne (Celtic) From the house of stone. #2

Trent (Latin) Torrent. #5

Trenton (Old English) Town on the river Trent. #7

Trevar (Gaelic) Wise and discreet. #3

Trever (Gaelic) Wise and discreet. #7

Trevin (Welsh) Fair town. #7

Trevor (Gaelic) Wise and discreet. #8

Trey (Old English) Three, the third. #5

Trip (Old English) To dance or hop. #9

Tripp (Old English) To dance or hop. #7

Tripper (Old English) To dance or hop. #3

Tristam (Welsh) Sorrowful. #1

Tristram (Welsh) Sorrowful. #1

Troy (Gaelic) Foot soldier. #6

True (Old English) Faithful man. #1

Trueman (Old English) Faithful man. #2

Truemann (Old English) Faithful man. #7

Trumaine (Old English) Faithful man. #2

Truman (Old English) Faithful man. #6

Tucker (Old English) Tucker of cloth. #6

Tully (Gaelic) He who lives with God's peace. #9

Turner (33/6) (Latin) Lathe worker. #6

Tyler (Old English) Maker of tiles. #8

Tylor (Old English) Maker of tiles. #9

Tynam (Gaelic) Dark. #1

Tynan (Gaelic) Dark. #2

Tyrone (Gaelic) Land of Owen. #7

Tyrus (22/4) (Old Norse) Thunder. #4

Tyson (Old French) Firebrand. #3

Udell (Old English) Prosperous. #9

Ugo (Nigerian) Eagle. #7

Ulric (Old German) Wolf-ruler. #9

Ulrich (Old German) Wolf-ruler. #8

Ulrick (Old German) Wolf-ruler. #2

Ulysses (Latin-Greek) Wrathful. #3

Upton (Old English) From the upper town. #5

Urbain (Latin) From the city. #2

Urban (Latin) From the city. #2

Urbano (Latin) From the city. #8

Urbanus (Latin) From the city. #6

Uriah (Hebrew) Jehovah is my light. #3

Uriel (Hebrew) God is my flame. #2

Urson (Latin) Little bear. #6

Valentin (Latin) Strong. #7

Valentine (Latin) Strong. #3

Valentino (Latin) Strong. #4

Van (Dutch) Of noble descent. #1

Vance (Old English) Thresher. #9

Vasilis (Greek) Magnificent knight. #1

Vaughan (Welsh) Small. #2

Vaughn (Welsh) Small. #1

Verdell (33/6) (Latin) Green, flourishing. #6

Vergil (Latin) Staff bearer. #1

Vern (Latin) Springlike, youthful. #5

Verne (Latin) Springlike, youthful. #1

Vernon (Latin) Springlike, youthful. #7

Victor (33/6) (Latin) Victorious. #6

Victorio (Latin) Victorious. #3

Vidal (Latin) Life. #3

Vincent (33/6) (Latin) Victorious. #6

Vincente (Latin) Victorious. #2

Vinnie (Latin) Victorious. #1

Vinny (Latin) Victorious. #3

Vinson (Old English) The conqueror's son. #3

Virgil (Latin) Staff-bearer. #5

Virgilio (Latin) Staff-bearer. #2

Vito (Latin) Alive. #3

Vittorio (Latin) Victorious. #2

Vladamir (Old Slavic) Powerful prince. #8

Vladimir (Old Slavic) Powerful prince. #7

Wade (Old English) Advancer. #6

Wainwright (Old English) Wagon-maker. #6

Waite (22/4) (Old English) Guard. #4

Wakefield (Old English) From the wet field. #4

Wald (Old German) Ruler. #4

Waldemar (Old German) Ruler. #5

Walden (Old English) From the woods. #5

Waldo (Old German) Ruler. #1

Waldon (Old English) From the woods. #6

Waldron (33/6) (Old German) Strength of the raven. #6

Walker (Old English) Thickener of cloth. #7

Walt (11/2) (Old German) Peaceful warrior. #2

Walter (Old German) Powerful warrior. #7

Walton (22/4) (Old English) From the walled town. #4

Ward (Old English) Guardian. #1

Wardell (Old English) Guardian. #3

Ware (Old German) Watchman or defender. #2

Waring (Old German) Watchman or defender. #9

Warner (Old German) Defending army or warrior. #7

Warren (Old German) Watchman or defender. #7

Washington (Old English) From the astute one's estate. #4

Wat (Old German) Mighty warrior. #8

Watford (33/6) (Old English) From the hurdle by the ford. #6

Waverley (Old English) Quaking aspen tree meadow. #3

Waverly (Old English) Quaking aspen tree meadow. #7

Waylan (22/4) (Old English) From the land by the road. #4

Wayland (Old English) From the land by the road. #8

Waylen (Old English) From the land by the road. #8

Waylon (Old English) From the land by the road. #9

Wayne (Old English) Wagoner. #5

Webb (Old English) Weaver. #5

Webster (Old English) Weaver. #2

Welby (22/4) (Old German) From the farm by the spring. #4

Weldon (Old English) From the well on the hill. #1

Wells (Old English) From the springs. #8

Wendall (Old German) The wanderer. #8

Wendel (Old German) The wanderer. #9

Wendell (Old German) The wanderer. #3

Werner (Old German)

Defending army or warrior. #2

Wernher (Old German) Defending army or warrior. #1

Wes (11/2) (Old English) From the western meadow. #2

Westbrook (Old English) From the western brook. #2

Westley (Old English) From the western meadow. #1

Weston (Old English) From the western estate. #6

Wheeler (Old English) Wheel-maker. #4

Whitaker (Old English) From the white field. #5

Whitby (33/6) (Old English) From the white field. #6

Whitman (Old English) White-haired man. #7

Whittaker (Old English) From the white field. #7

Wiatt (Old French) Little warrior. #1

Wilber (33/6) (German-French) Brilliant hostage. #6

Wilbert (German-French) Brilliant hostage. #8

Wilbur (German-French) Brilliant hostage. #4

Wilburt (33/6) (German-French) Brilliant hostage. #6

Wilden (Old English) From the wooded hill. #4

Wildon (Old English) From the wooded hill. #5

Wiley (Old English) From the water meadow. #2

Wilfred (Old German) Resolute and peaceful. #5

Wilfredo (Old German) Resolute and peaceful. #2

Wilfrid (Old German) Resolute and peaceful. #9

Wilhelm (Old German) Determined guardian. #1

Wilkie (33/6) (Old German) Determined guardian. #6

Will (Old German) Determined guardian. #2

Willard (Old German) Resolutely brave. #7

William (Old German) Determined guardian. #7

Willis (Old German) Determined guardian. #3

Willy (Old German) Determined guardian. #9

Wilmer (Old English) Determined guardian. #8

Wilmus (Old German) Determined guardian. #7

Wilson (Old English) Son of Will. #2

Wilt (Old English) From the town by the spring. #1

Wilton (Old English) From the town by the spring. #3

Win (Old English) From the friendly place. #1

Windham (Scottish Gaelic) Village near the winding road. #9

Winfield (Old English) From the friendly field. #1

Winn (Old English) From the friendly place. #6

Winslow (Old English) From the friend's hill. #7

Winston (33/6) (Old English) From the friend's town. #6

Winthrop (Old English) Dweller at the friend's estate. #6

Wolf (Old German) Wolf. #2

Wolfgang (Old German) Advancing wolf. #4

Wolfie (Old German) Wolf. #7

Wolfram (Old German) Respected and feared. #7

Wolfy (Old German) Wolf. #9

Wood (Old English) From the passage in the woods. #3

Woodman (Old English) From the passage in the woods. #4

Woodrow (Old English) From the passage in the woods. #5

Woody (Old English) From the passage in the woods. #1

Woolsey (33/6) (Old German) Victorious wolf. #6

Worth (Old English) From the farmstead. #3

Worthington (Old English) From the farmstead. #1

Worthy (Old English) From the farmstead. #1

Wright (Old English) Carpenter. #4

Wyatt (Old French) Little warrior. #8

Wylie (Old English) From the water meadow. #2

Wyndham (Scottish Gaelic) Village near the winding road. #7

Wynn (22/4) (Old English) From the friendly place. #4

Xanthus (Latin) Golden-haired. #8

Xaver (Arabic) Bright. #7

Xavier (Arabic) Bright. #7

Xeno (22/4) (Greek) Strange voice. #4

Xenophon (Greek) Strange voice. #3

Xenos (Greek) Stranger. #5

Xerxes (Persian) Ruler. #5

Ximenes (Hebrew) One who hears. #8

Ximenez (Hebrew) One who hears. #6

Xiomar (Old German) Famous in battle. #8

Xylon (Greek) From the forest. #9

Yale (Old English) From the corner of the land. #7

Yancy (American Indian) Englishman. #5

Yardley (Old English) From the enclosed meadow. #9

Yates (Old English) Dweller at the gates. #7

Yehudi (Hebrew) Praise the Lord. #9

York (Celtic) Yew-tree estate. #6

Yule (Old English) Born at Yuletide. #9

Yules (Greek) Youthful, downy-bearded. #1

Yuma (American Indian) Son of a chief. #6

Yuri (Hebrew) Jehovah is my light. #1

Yuriko (Hebrew) Jehovah is my light. #9

Yves (Norse) Battle archer. #8

Yvon (22/4) (Old French) Archer with a yew bow. #4

Zaccheus (Hebrew) Jehovah has remembered. #5

Zachariah (Hebrew) Jehovah has remembered. #3

Zacharias (Hebrew) Jehovah has remembered. #5

Zachary (Hebrew) Jehovah has remembered. #1

Zachery (Hebrew) Jehovah has remembered. #5

Zander (Greek) Helper of humankind. #5

Zane (Hebrew) God's gracious gift. #1

Zared (Hebrew) Ambush. #9

Zarek (Slavic) God protects. #7

Zeb (Hebrew) The Lord's gift. #6

Zebadiah (Hebrew) The Lord's gift. #2

Zebulen (Hebrew) Dwelling place. #4

Zebulon (Hebrew) Dwelling place. #5

Zedekiah (Hebrew) God is mighty and just. #6

Zeke (Hebrew) Strength of God. #2

Zelig (Old German) Blessed. #5

Zeno (Greek) Life. #6

Zephaniah (Hebrew) Treasured by the Lord. #7

Zeus (Greek) Living. #8

MASTER LIST—UNISEX NAMES

Abbe (Hebrew) My father is joy. #1

Abbey (Hebrew) My father is joy. #8

Abbie (Hebrew) My father is joy. #1

Abby (Hebrew) My father is joy. #3

Adriaan (Latin) Black earth. #3

Adrian (Latin) Black earth. #2

Alex (Greek) Helper of humankind. #6

Alexi (Greek) Helper of humankind. #6

Alexie (Greek) Helper of humankind. #2

Alexis (Greek) Helper of humankind. #7

Ali (Arabic) Greatest. #4

Allie (Greek) Truthful. #3

Andy (Greek) Strong and manly. #8

Angel (Greek) Heavenly messenger. #3

Angie (Greek) Heavenly messenger. #9

Ardell (Latin) Ardent, fiery. #7

Ardelle (Latin) Ardent, fiery. #3

Arden (Latin) Ardent, fiery. #6

Ariel (Hebrew) Lioness of God. #9

Arnelle (Old German) Strong as an eagle. #4

Arnett (Old German) Little eagle. #6

Ash (Old English) Ash tree. #1

Ashley (Old English) From the ash tree meadow. #7

Ashton (Old English) From the ash tree town. #5

Aubrey (Old French) Blond ruler. #9

Aubry (22/4) (Old French) Blond ruler. #4

Austen (Latin) Majestic dignity. #8

Austin (Latin) Majestic dignity. #3

Avery (Old English) Good counselor. #8

Bailey (Old French) Bailiff, steward. #9

Barrie (Celtic) One whose intellect is sharp. #8

Bethel (Hebrew) House of God. #7

Beverly (Old English) From the beaver meadow. #8

Billie (Old German) Determined guardian. #4

Billy (Old German) Determined guardian. #6

Blaine (Gaelic) Thin, lean. #7

Blair (Gaelic) From the plain. #6

Blaire (Gaelic) From the plain. #2

Blaise (Latin) Stammerer. #3

Blake (Old English) Dark hair and complexion. #4

Blane (Gaelic) Thin, lean. #7

Blayne (Gaelic) Thin, lean. #5

Blaze (Latin) Stammerer. #1

Bobbi (Old English) Bright fame. #3

Bobbie (Old English) Bright fame. #8

Bobby (Old English) Bright fame. #1

Brandee (Dutch) Brandy drink. #4

Brandi (Dutch) Brandy drink. #3

Brandy (Dutch) Brandy drink. #1

Brett (Celtic) Native of Brittany. #2

Britt (Latin) From England. #6

Brook (Old English) From the brook. #7

Brooke (Old English) From the brook. #3

Brooks (Old English) From the brook. #8

Cace (Gaelic) Vigilant. #3

Caley (Gaelic) Thin, slender. #1

Cameron (33/6) (Scottish Gaelic) Crooked nose. #6

Carey (Old German) Strong and masculine. #7

Carmel (Hebrew) God's vineyard. #7

Carol (22/4) (Old German) Noble and strong. #4

Carroll (Old German) Noble and strong. #7

Cary (Old German) Strong and masculine. #2

Case (Gaelic) Vigilant. #1

Casey (Gaelic) Vigilant. #8

Cass (Gaelic) Clever. #6

Cassidy (Gaelic) Clever. #8

Chandler (Old French) Candlemaker. #2

Channing (Old English) Knowing. #7

Charlie (Old German) Strong and masculine. #2

Chris (Greek) Anointed. #3

Clare (Latin) Bright, shining girl. #3

Codi (22/4) (Old English) A cushion. #4

Cody (Old English) A cushion. #2

Connie (33/6) (Latin) Firmness, constancy. #6

Cord (22/4) (Old French) Ropemaker. #4

Corey (Gaelic) From the hollow. #3

Cortney (Old English) From the court. #1

Cory (Gaelic) From the hollow. #7

Courtenay (Old English) From the court. #5

Courtney (Old English) From the court. #4

Cruz (Spanish) Cross. #5

Cyd (Old French) From Saint Denis. #5

Cydney (Old French) From Saint Denis. #4

Dagny (Scandinavian) Day, brightness. #6

Dakota (American Indian) Friend, ally. #7

Dale (Old English) From the valley. #4

Dallas (Gaelic) Wise. #4

Dana (11/2) (Old English) From Denmark. #2

Daniel (Hebrew) God is my judge. #9

Danny (22/4) (Hebrew) God is my judge. #4

Darby (Gaelic) Free man. #5

Darcie (Old French) From the fortress. #4

Darcy (Old French) From the fortress. #6

Darel (22/4) (Old French) Beloved. #4

Daren (Old French) From the fortress. #6

Darin (Gaelic) Great. #1

Daron (Gaelic) Great. #7

Darren (33/6) (Gaelic) Great. #6

Darryl (33/6) (Old French) Beloved. #6

Daryl (Old French) Beloved. #6

Daryn (Gaelic) Great. #8

Dennie (33/6) (Greek) God of wine. #6

Denny (Greek) God of wine. #8

Devan (Old English) Defender of Devonshire. #1

Devin (Gaelic) Poet. #9

Diamond (33/6) (Greek) Of high value, brilliant. #6

Doran (Celtic) Stands firm. #7

Dore (Greek) Gift. #6

Dorian (Greek) From the sea. #7

Drew (Latin) Strong. #5

Dru (Latin) Strong. #7

Dusty (Old German) Valiant fighter. #8

Eden (Hebrew) Delight. #1

Erin (Gaelic) Peace. #1

Essex (Old English) From the east. #9

Evelin (Hebrew) Life-giving. #4

Evelyn (Hebrew) Life-giving. #2

Flo (Latin) Flowering or blooming. #6

Florence (Latin) Flowering or blooming. #6

Fortune (Latin) Fortunate. #9

Fran (Latin) Free. #3

Francis (Latin) Free. #7

Frankie (Latin) Free. #1

Franky (Latin) Free. #3

Freddie (Old German) Peaceful ruler. #6

Freddy (Old German) Peaceful ruler. #8

Gabriel (Hebrew) Devoted to God. #9

Gaby (Hebrew) Devoted to God. #8

Gael (Old English) Gay, lively. #7

Gail (Old English) Gay, lively. #2

Gale (Old English) Gay, lively. #7

Garland (Old French) Wreath. #3

Garnet (Old English) Garnet gem. #2

Garnett (Old English) Garnet gem. #4

Gene (22/4) (Greek) Wellborn noble. #4

Georgie (Greek) Landholder, farmer. #3

Geri (Old German) Spear-mighty. #3

Germain (Old German) People of the spear. #4

Germaine (Old German) People of the spear. #9

Gerrie (Old German) Spear-mighty. #8

Gerry (Old German) Spear-mighty. #1

Glen (Old Welsh) Dweller in a valley. #2

Glenn (Old Welsh) Dweller in a valley. #7

Guadalupe (Arabic) River of black stones. #7

Hadley (Old English) From the heath. #1

Hailey (33/6) (Old English) Hay meadow. #6

Haleigh (Old English) Hay meadow. #5

Haley (Old English) Hay meadow. #6

Halley (Old English) Hay meadow. #9

Hallie (Old English) Hay meadow. #2

Harley (33/6) (Old English) From the long field. #6

Haven (Old English) Place of safety. #5

Hayatt (Old English) From the high gate. #3

Hayley (Old English) Hay meadow. #4

Hilary (Latin) Cheerful. #1

Hillary (Latin) Cheerful. #4

Hollis (Old English) From the grove of holly trees. #3

Holly (Old English) Holly tree. #9

Jackie (Hebrew) The supplanter. #3

Jacky (Hebrew) The supplanter. #5

Jamie (Hebrew) The supplanter. #2

Jan (Hebrew) God's gift of grace. #7

Jaye (Old French) Blue jay. #5

Jean (Hebrew) God's gift of grace. #3

Jerry (Latin) Sacred or holy name. #4

Jess (Hebrew) Wealthy. #8

Jesse (Hebrew) Wealthy. #4

Jessie (22/4) (Hebrew) Wealthy. #4

Jessy (Hebrew) Wealthy. #6

Joan (Hebrew) God's gift of grace. #4

Jodie (Hebrew) Admired. #7

Jody (Hebrew) Admired. #9

Johnny (Hebrew) God's gracious gift. #5

Jordan (Hebrew) Downflowing. #8

Julie (Greek) Young in heart and mind. #3

Kacey (Gaelic) Vigilant. #9

Kacie (Gaelic) Vigilant. #2

Kai (Old Welsh) Keeper of the keys. #3

Karol (Old German) Noble and strong. #3

Kasey (Gaelic) Vigilant. #7

Kassidy (Gaelic) Clever. #7

Kellen (Gaelic) Warrior. #5

Kelley (Gaelic) Warrior. #7

Kelly (Gaelic) Warrior. #2

Kelsey (Old Norse) From the ship island. #5

Kendall (Old English) From the bright valley. #5

Kerry (Gaelic) Dark one. #5

Kiki (22/4) (Old German) Lord of the manor. #4

Kiley (Gaelic) Handsome. #8

Kim (Old English) From the king's wood. #6

Kimberly (Old English) From the king's wood. #5

Kip (Old English) From the pointed hill. #9

Kipp (Old English) From the pointed hill. #7

Kirby (Old Norse) From the church village. #2

Kit (Greek) Bearer of Christ. #4

Kodi (Old English) A cushion. #3

Kody (Old English) A cushion. #1

Korey (Gaelic) From the hollow. #2

Kory (Gaelic) From the hollow. #6

Kris (Greek) Anointed. #3

Kristin (Greek) Anointed. #1

Lane (Old English) From the narrow road. #5

Lanie (Old English) From the narrow road. #5

Lanny (Old English) From the narrow road. #3

Lauren (Latin) Crowned with laurels. #8

Laurie (Latin) Crowned with laurels. #3

Lee (Old English) From the meadow. #4

Leigh (Old English) From the meadow. #5

Lesley (Scottish Gaelic) From the grey fortress. #6

Leslie (Scottish Gaelic) From the grey fortress. #8

Linden (Old English) From the linden tree. #4

Lindsay (Old English) From the linden tree island. #3

Lindsey (Old English) From the linden tree island. #7

Linsey (Old English) From the linden tree island. #3

Lisle (Old French) From the island. #3

Loni (Old English) Solitary. #5

Lonnie (33/6) (Old English) Solitary. #6

Lonny (Old English) Solitary. #8

Loren (Latin) Crowned with laurels. #1

Lorin (Latin) Crowned with laurels. #5

Lou (Old German) Famous warrior. #3

Lupe (Arabic) River of black stones. #9

Lyall (Old French) From the island. #8

Lyell (Old French) From the island. #3

Lyn (Old English) From the waterfalls. #6

Lyndsay (Old English) From the linden tree island. #1

Lyndsey (Old English) From the linden tree island. #5

Lynn (Old English) From the waterfalls. #2

Mackenzie (Gaelic) Son of the wise leader. #6

Macy (Old French) Mace. #6

Maddy (Hebrew) Gift of God. #2

Madison (Old English) Son of the powerful soldier. #3

Mallory (33/6) (Old German) Army counselor. #6

Malory (Old German) Army counselor. #3

Marion (Hebrew) Bitter, graceful. #7

Marlen (Old English) Falcon or hawk. #9

Marlin (Old English) Falcon or hawk. #4

Marlo (Old English) From the hill by the lake. #5

Marlon (Old English) Falcon or hawk. #1

Marlow (Old English) From the hill by the lake. #1

Marlowe (33/6) (Old English) From the hill by the lake. #6

Marty (Latin) Follower of Mars. #5

Mattie (Hebrew) Gift of God. #5

Matty (Hebrew) Gift of God. #7

Maxie (Latin) Most excellent. #7

Maxy (Latin) Most excellent. #9

McKenzie (Gaelic) Son of the wise leader. #5

Mead (Old English) From the meadow. #5

Meade (Old English) From the meadow. #1

Mel (Gaelic) Brilliant chief. #3

Meredith (Old Welsh) Guardian from the sea. #1

Meridith (Old Welsh) Guardian from the sea. #5

Merle (French) Blackbird. #8

Micah (Hebrew) Like unto the lord. #7

Michael (33/6) (Hebrew) Like unto the Lord. #6

Michel (Hebrew) Like unto the Lord. #5

Micky (Hebrew) Like unto the Lord. #7

Montana (Latin) Mountain. #6

Morgan (Scottish Gaelic) From the edge of the sea. #5

Morgen (Scottish Gaelic) From the edge of the sea. #9

Murphy (Gaelic) Sea warrior. #2

Nealy (Gaelic) The champion. #3

Neddy (Old English) Rich and happy protector. #7

Neely (Gaelic) The champion. #7

Nevada (Spanish) Snow-clad. #2

Nickie (33/6) (Greek) The people's victory. #6

Nicky (Greek) The people's victory. #8

Nicol (Greek) The people's victory. #8

Nika (Greek) The people's victory. #8

Nike (Greek) The people's victory. #3

Niki (Greek) The people's victory. #7

Nikki (Greek) The people's victory. #9

Noel (French) Born at Christmas. #1

Noele (French) Born at Christmas. #6

Pace (Latin) Born at Easter or Passover. #7

Padget (French) Useful assistant. #8

Padgett (French) Useful assistant. #1

Page (French) Useful assistant. #2

Paige (French) Useful assistant. #2

Palm (Latin) The palm-bearer. #6

Palmer (Latin) The palm-bearer. #2

Pat (Latin) Wellborn, noble. #1

Patsy (Latin) Wellborn, noble. #9

Patty (Latin) Wellborn, noble. #1

Paule (Latin) Little. #1

Pepi (Hebrew) He shall add. #1

Phoenix (Greek) The eagle risen from the ashes. #1

Quin (Gaelic) Wise, intelligent. #7

Quinn (Gaelic) Wise, intelligent. #3

Rabi (Arabic) Breeze. #3

Rande (Old English) Swift wolf. #6

Randee (Old English) Swift wolf. #2

Randi (Old English) Swift wolf. #1

Randy (Old English) Swift wolf. #8

Ray (Old French) Kingly. #8

Reagan (Gaelic) Little king. #1

Reagen (Gaelic) Little king. #5

Rebel (Latin) Opposer, revolutionary. #6

Regan (Gaelic) Little king. #9

Reggie (Old English) Mighty and powerful. #6

Rene (Latin) Reborn. #6

Rennie (Latin) Reborn. #2

Renny (Latin) Reborn. #4

Rickie (Old German) Wealthy, powerful. #1

Ricky (Old German) Wealthy, powerful. #3

Riki (Old German) Wealthy, powerful. #2

Rikki (Old German) Wealthy, powerful. #4

Riley (33/6) (Gaelic) Valiant. #6

Robbie (33/6) (Old English) Bright fame. #6

Robby (Old English) Bright fame. #8

Robin (Old English) Bright fame. #4

Robyn ((Old English) Bright fame. #2

Rockey (Old English) From the rock. #5

Rocky (Old English) From the rock. #9

Rodie (33/6) (Latin) Dew of the sea. #6

Rody (Latin) Dew of the sea. #8

Ronnie (Old English) Mighty and powerful. #3

Ronny (Old English) Mighty and powerful. #5

Rosario (Spanish) Rosary. #5

Rowan (Gaelic) Red-haired. #8

Rowe (Gaelic) Red-haired. #7

Rowen (Gaelic) Red-haired. #3

Ruby (Old French) Ruby gemstone. #3

Rusty (22/4) (French) Red-haired. #4

Ryan (22/4) (Gaelic) Little king. #4

Ryann (Gaelic) Little king. #9

Ryley (Gaelic) Valiant. #4

Ryon (Gaelic) Little king. #9

Ryun (Gaelic) Little king. #6

Sacha (Greek) Helper of humankind. #5

Sal (Italian) Savior. #5

Sammy (Hebrew) His name is God. #8

Sandie (Old English) From the sandy hill. #7

Sandy (Old English) From the sandy hill. #9

Sandye (Old English) From the sandy hill. #5

Sasha (Greek) Helper of humankind. #3

Scottie (Old English) Scotsman. #1

Scotty (Old English) Scotsman. #3

Sean (Hebrew) God's gracious gift. #3

Shandy (Old English) Rambunctious. #8

Shane (Hebrew) God's gracious gift. #2

Shannon (Gaelic) Small and wise. #4

Shawn (Hebrew) God's gracious gift. #2

Shay (Gaelic) From the fairy fort. #8

Shea (Gaelic) From the fairy fort. #6

Shelby (Old English) From the estate on the ledge. #8

Shelley (Old English) From the meadow on the ledge. #5

Shelly (Old English) From the meadow on the ledge. #9

Sheridan (Celtic) Wild man or savage. #6

Sidnee (Old French) From Saint Denis. #2

Sidney (Old French) From Saint Denis. #4

Simeon (Hebrew) One who hears. #3

Sky (English) Sky. #1

Skye (Dutch) Sheltering. #6

Skylar (Dutch) Scholar. #5

Skyler (Dutch) Scholar. #9

Sloan (Gaelic) Warrior. #7

Sloane (Gaelic) Warrior. #3

Stace (Latin) Stable, prosperous. #3

Stacey (Latin) Stable, prosperous. #1

Stacie (Latin) Stable, prosperous. #3

Stacy (Latin) Stable, prosperous. #5

Stevie (Greek) Crowned. #8

Stevy (Greek) Crowned. #1

Storm (22/4) (English) Storm. #4

Stormy (English) Stormy. #2

Sydney (Old French) From Saint Denis. #2

Tailor (Old English) Tailor. #3

Taran (Gaelic) Rocky pinnacle. #9

Taryn (Gaelic) Rocky pinnacle. #6

Tate (Old English) Cheerful. #1

Taylor (Old English) Tailor. #1

Teddi (Greek) Gift of God. #6

Teddy (22/4) (Greek) Gift of God. #4

Temple (Old English) From the town of the temple. #8

Terri (Greek) The harvester. #7

Terry (Greek) The harvester. #5

Tierney (Gaelic) Lordly one. #6

Tobe (Hebrew) The Lord is good. #6

Tobey (22/4) (Hebrew) The Lord is good. #4

Toby (Hebrew) The Lord is good. #8

Tommie (Hebrew) The devoted brother. #3

Tonnie (Latin) Praiseworthy, without a peer. #5

Tony (Latin) Praiseworthy, without a peer. #2

Tracee (Gaelic) Battler. #7

Tracey (Gaelic) Battler. #9

Tracie (Gaelic) Battler. #2

Tracy (22/4) (Gaelic) Battler. #4

Tristan (Welsh) Sorrowful. #2

Vail (Old English) From the valley. #8

Val (Old English) From the valley. #8
Vale (Old English) From the valley. #4
Valle (Old English) From the valley. #7
Vallie (Old English) From the valley. #7
Vally (Old English) From the valley. #9
Vann (Dutch) Of noble descent. #6
Ventura (Spanish) Happiness, good luck. #2
Vivian (Latin) Alive. #5
Vivien (Latin) Alive. #9

Wallace (Old English) From Wales. #3
Wally (Old German) Powerful warrior. #1
Weslee (Old English) From the western meadow. #6
Wesley (Old English) From the western meadow. #8
Whitley (Old English) From the white field. #3
Whitney (Old English) From the white island. #5
Willie (Old German) Determined guardian. #7
Winny (Old German) Peaceful friend. #4

Appendix 2

Lucky Colors, Gems, and Metals

Number	Planet	Element	Color	Gemstone	Metal
1	Sun	Fire	Gold Orange Yellow	Amber Topaz Citrine	Gold
2	Moon	Water	Cream White Green	Pearl Moonstone Jade	Silver
3	Jupiter	Fire	Violet Mauve Lilac	Amethyst Garnet	Tin
4	Uranus	Earth	Blue Grey Plaid	Sapphire Topaz	Uranium
5	Mercury	Air	Grey Silver	Diamond	Quicksilver

Number	Planet	Element	Color	Gemstone	Metal
			White		
6	Venus	Earth	Blue Rose Pink	Turquoise Emerald	Copper
7	Neptune	Water	Green Pastels White	Jade Moonstone Cat's Eye	Uranium
8	Saturn	Earth	Purple Black Brown	Sapphire Black Pearl Black Diamond	Lead
9	Mars	Fire	Red Rose Pink	Ruby Garnet Bloodstone	Iron

Appendix 3

PYTHAGOREAN SYSTEM
NUMBER/LETTER CORRESPONDENCES CHART

1	2	3	4	5	6	7	8	9	11	22	33
A	B	C	D	E	F	G	H	I			
J	K	L	M	N	O	P	Q	R	(K)		
S	T	U	V	W	X	Y	Z			(V)	(X)

Worksheets

	Names	Totals
First Name _____		
Numbers _____		
Reduced _____		
Middle Name _____		
Numbers _____		
Reduced _____		
Last Name _____		
Numbers _____		
Reduced _____		
Complete Name _____		
Reduced _____		
Complete Name Total _____		

	Names	Totals
First Name _____		
Numbers _____		
Reduced _____		
Middle Name _____		
Numbers _____		
Reduced _____		
Last Name _____		
Numbers _____		
Reduced _____		
Complete Name _____		
Reduced _____		
Complete Name Total _____		

Worksheets

	Names	Totals
First Name _____		
Numbers _____		
Reduced _____		
Middle Name _____		
Numbers _____		
Reduced _____		
Last Name _____		
Numbers _____		
Reduced _____		
Complete Name _____		
Reduced _____		
Complete Name Total _____		

	Names	Totals
First Name _____		
Numbers _____		
Reduced _____		
Middle Name _____		
Numbers _____		
Reduced _____		
Last Name _____		
Numbers _____		
Reduced _____		
Complete Name _____		
Reduced _____		
Complete Name Total _____		

Worksheets

	Names	Totals
First Name	_____	
Numbers	_____	
Reduced	_____	
Middle Name	_____	
Numbers	_____	
Reduced	_____	
Last Name	_____	
Numbers	_____	
Reduced	_____	
Complete Name	_____	
Reduced	_____	
Complete Name Total	_____	

	Names	Totals
First Name	_____	
Numbers	_____	
Reduced	_____	
Middle Name	_____	
Numbers	_____	
Reduced	_____	
Last Name	_____	
Numbers	_____	
Reduced	_____	
Complete Name	_____	
Reduced	_____	
Complete Name Total	_____	

Worksheets

	Names	Totals
First Name	_____	_____
Numbers	_____	_____
Reduced	_____	_____
Middle Name	_____	_____
Numbers	_____	_____
Reduced	_____	_____
Last Name	_____	_____
Numbers	_____	_____
Reduced	_____	_____
Complete Name	_____	_____
Reduced	_____	_____
Complete Name Total	_____	_____

	Names	Totals
First Name	_____	_____
Numbers	_____	_____
Reduced	_____	_____
Middle Name	_____	_____
Numbers	_____	_____
Reduced	_____	_____
Last Name	_____	_____
Numbers	_____	_____
Reduced	_____	_____
Complete Name	_____	_____
Reduced	_____	_____
Complete Name Total	_____	_____

Worksheets

	Names	Totals
First Name	_____	
Numbers	_____	
Reduced	_____	
Middle Name	_____	
Numbers	_____	
Reduced	_____	
Last Name	_____	
Numbers	_____	
Reduced	_____	
Complete Name	_____	
Reduced	_____	
Complete Name Total	_____	

	Names	Totals
First Name	_____	
Numbers	_____	
Reduced	_____	
Middle Name	_____	
Numbers	_____	
Reduced	_____	
Last Name	_____	
Numbers	_____	
Reduced	_____	
Complete Name	_____	
Reduced	_____	
Complete Name Total	_____	

Printed in the United States
by Baker & Taylor Publisher Services